Coordinate Geometry

I0475325

Coordinate Geometry

Dedicated to my team

Coordinate Geometry

Copyright © 2011 by Chandramouli Mahadevan,
On behalf of Astrarka
All rights reserved.
No part of this book may be reproduced, stored, or transmitted by any means, whether auditory, graphic, mechanical, or electronic, without written permission of both publisher and author, except in the case of brief excerpts used in critical articles and reviews. Unauthorized reproduction of any part of this work is illegal and is punishable by law.

ISBN-13: 978-1466327214
ISBN-10: 1466327219
First Edition 2011

Foreword

I have often looked for inspiration from old proverbs and pithy statements. For example, "If you do not know where to go, any road would do." – might be a good way to remember that there can be infinite straight lines passing through a single point. The "where to go" question is answered by the definition of a second point. If you know the starting point and the end point, you can draw one straight line lying on a plane. In general, it is important to know our current position and have a very good idea about where we are headed. If you do not know where you are, no map can help. This book deals with maps, paths and trajectories – in a matter of speaking. Welcome to the world of Coordinate Geometry.

In classical mathematics, analytic geometry, also known **as Coordinate Geometry**, or **Cartesian geometry**, is the study of geometry using a coordinate system and the principles of algebra and analysis. This contrasts with the synthetic approach of Euclidean geometry. In Euclidean Geometry, we defined shapes and concepts – proved the theorem through logical reasoning built on Axioms and other theorems. An axiom denotes primitive statements that are assumed to be true. Theorems are statements that can be proved purely by using axioms and other theorems facts. This is also known as deductive reasoning; and you can see wide applications of this problem solving paradigm – in topics such as symbolic computation and predicate calculus. Coordinate geometry is widely used in physics and engineering, and is the foundation of most modern fields of geometry, including algebraic, differential, discrete, and computational geometry.

Usually the Cartesian coordinate system is applied to manipulate equations for planes, straight lines, and squares, often in two and sometimes in three dimensions of measurement. Geometrically, one studies the Euclidean plane (2 dimensions) and Euclidean space (3 dimensions). We will be looking at 2-dimensional coordinate geometry. As taught in school books, analytic geometry can be explained more simply: it is concerned with defining and representing geometrical shapes in a numerical way and extracting numerical information from shapes' numerical definitions and representations. The numerical output, however, might also be a vector or a shape.

It is believed that the Greek mathematician Menaechmus solved problems and proved theorems by using a method that had a strong resemblance to the use of coordinates and it has sometimes been maintained that he had introduced analytic geometry. Apollonius of Perga, in On Determinate Section, dealt with problems in a manner that may be called an analytic geometry of one dimension; with the question of finding points on a line that were in a ratio to the others. Apollonius in the Conics further developed a method that is so similar to analytic geometry that his work is sometimes thought to have anticipated the work of Descartes — by some 1800 years. His application of reference lines, a diameter and a tangent is essentially no different than our modern use of a coordinate frame, where the distances measured along the diameter from the point of tangency are the abscissas, and the segments parallel to the tangent and intercepted between the axis and the curve are the ordinates. He further developed relations between the abscissas and the corresponding ordinates that are equivalent to rhetorical equations of curves. However, although Apollonius came close to developing analytic geometry, he did not manage to do so since he did not take into account negative magnitudes and in every case the coordinate system was superimposed upon a given curve a posteriori instead of a priori. That is, equations were determined by curves, but curves were not determined by equations. Coordinates, variables, and equations were subsidiary notions applied to a specific geometric situation. The growth of this field continued until the modern times.

We will be taking a look at the building blocks and concepts – and solve a few problems to give a flavor of problem solving process. A detailed look at problem solving is left to the sequel of this book – "Problems in Coordinate Geometry".

Chandramouli Mahadevan

Astrarka.

Bangalore, India.

Preface

This book is an integral part of a series on Coordinate Geometry.

The book "Coordinate Geometry" covers the concepts involved in the various topics of this subject. A few selected problems are solved after each chapter, to aid the understanding of the student. The book finishes with a collection of problems that the student must practice on, to gain expertise.

"Problems in Coordinate Geometry" is a comprehensive solution set to the battery of over 700 problems in all topics covers in the first volume. The student is expected to make an honest attempt to solve the problems before looking at the suggested solutions. These solutions are systematic and comprehensive. No intermediate steps are skipped; which ensures that the overall flow of the problem solving process starting with the initial conditions to the final solution is maintained.

The best way to use this book is for the student to attempt each problem on his/her own. In doing so, the depth of understanding in the subject improves. Mathematics is not a spectator sport. It requires patience, perseverance and practice. The level of expertise in the subject in some sense is directly proprotional to the number of problems solved by the student. The term "solved" is used to imply accuracy of thought, stringing together intermediate steps and accuracy of the final result. In a way, this term refers to the quality of the means and the quality of the end goal for each problem.

This work is a comprehensive self study guide for the students who desire to improve their understanding, appearing for Mathematics related competitive examinations and tests. These works are based on the gold standard on the topic by SL Loney. They published the book in late 1800s. This forms the central reference in several schools and colleges across the globe.

I believe that Astrarka has been blessed to have had the opportunity to work with some of the best and brightest Any work of this magnitude is always a product of teamwork. R Balasubramanian, Shilpa Jaikumar and Venkatratnam Pandit have contributed a great deal to this effort. A big thanks goes to the family members of our team. They have been a great source of inspiration during this entire effort. They have made a personal sacrifice to ensure that Astrarka succeeds. Without the unflinching commitment and single minded dedication of my team and the members of their family, this book would have been an exercise in futility.

Chandramouli Mahadevan

Table of Contents

Coordinate Geometry

1 Introduction

To say that Coordinate Geometry is useful, therefore, we must learn it, is an understatement. This book focuses on problem solving strategies. We have organized the material into problems, the solution of each problem immediately after the statement. Familiarity with middle school arithmetics and elementary algebra is assumed.

This book must not be read like a work of fiction. Instead, the student is advised to spend quality time in ensuring conceptual understanding. Mathematics requires three skills. Let us look at these issues.

Comprehension: At the core of Mathematics, we see the underlying patterns and designs. Each little node in this web is intimately related to the others around it. It is this intricate web of concepts that we need to pay attention to. Expertise and love for the subject is directly related to the quality of our comprehension. Our confidence to deal with issues related to any domain of knowledge is related to the quality of comprehension. So, we need to pay attention to the details. Taking notes is a good way to demonstrate our understanding and reinforce our learnings.

Problem Solving: The key to problem solving is practice. Math is not a spectator sport. There are no brownie points for being armchair diplomats. We need to be prepared to jump in and solve the problems that we come across. With practice, and only with practice do we gain the expertise to deploy the right ammunition to crack a problem.

Goal Clarity: Solving problems in order to verify our conceptual understanding is extremely important. Most of us believe arriving at the final answer is the ultimate goal. We have come across several books on the subject, where the authors have skipped several steps and simply used the phrase "it follows from the fundamental principles ..." and made a conclusion. We disagree with this approach. The purpose of the problem solving is build the path to the solution using first principles or well-known formulas - and build an airtight reasoning on how the problem solving process moves towards the final answer.

This serves as a demonstration of our understanding of the subject - basics, formulas and methods of manipulation.

2. Building Blocks

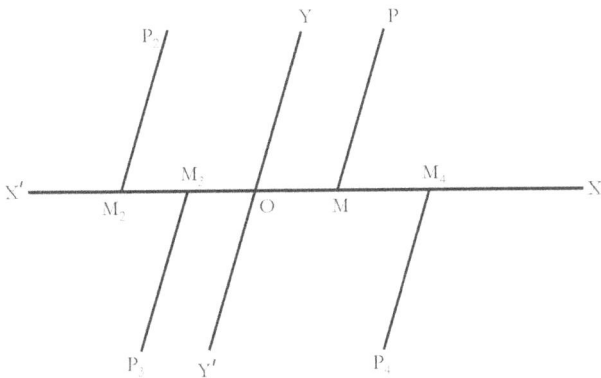

Fig. 1

1. Co-ordinates: This is the central theme or concept behind the content of this book. It is important for us to spend a few seconds to understand the term. When we talk of coordinates, we are talking about an unambiguous way to fix the position of an object or a thing with respect to a know reference point, object or a thing. When we say point A is 5 feet away from point B, it could any point which is 5 feet away from point B, in any direction. So, in order to disambiguate, we add the direction to the "left of point A" or "directly above point A" to complete the definition. Armed with this understanding, let us turn our attention to the mathematical notion of coordinates.

2. Let OX and OY be two fixed straight lines in the plane of paper. The line OX is called the X-axis, the line OY is called the Y-axis. Both are called the axes of co-ordinates.

3. The point O is called origin. This serves as the point of reference. Now, we can talk of a point P, which is 5 feet away from O, at an angle of 45^0 to line OX. This example helps us understand the need for direction to unambiguously define position; in addition to the "distance" from a known reference called "origin".

4. From any point P in the plane draw a straight line parallel to OY to meet OX in M. The distance OM is called the **abscissa** and the distance MP, the **ordinate** of the point P. Both measurements – abscissa and ordinate – constitute the co-ordinates of point P.

5. The distance measured parallel to OX is called x coordinate, and distance measured parallel to OY is called y coordinate. We frequently use suffixes like x_1, x_2, y_1, y_2 etc. to indicate several points or measurements. In the figure 1, $OM = x$ and $MP = y$. Therefore, we refer to coordinates of point P as (x, y).

6. Conversely, $P(x, y)$ is given, then we can determine its position from 'O'. To do this, we measure a distance $OM (= x)$ along OX and then from M measure a distance $MP (= y)$ parallel to OY. The point of intersection of these two lines is the required point $P(x, y)$.

7. Produce OX backwards to form the line OX' and OY backwards to become OY'. We use the convention that lines measured parallel to OX are positive; while those measured parallel to OX' are negative. Similarly, lines measured parallel to OY are positive and those parallel to OY' are negative.

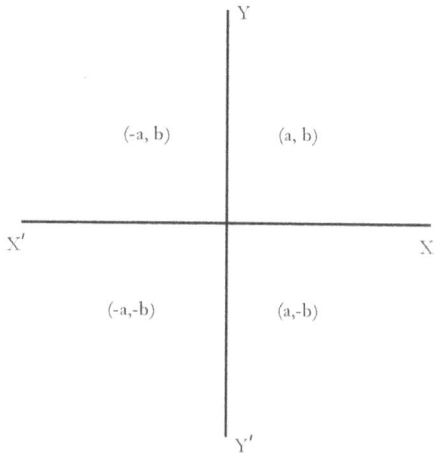

Fig. 2

8. In other words distances measured to the left of origin O are negative and those measured to the right of origin O are positive.Similarly, distances measured above the origin are positive and those measured below the origin are negative.

9. Thus we can split the coordinates into four quadrants.

    a. In the **first quadrant**, one to the right and above the origin, both x and y are positive. This is denoted by (a,b) in Fig 2.

    b. In the **second quadrant**, one to the left and above the origin, x is negative, while y is positive. This is denoted by $(-a,b)$ in Fig 2.

    c. In the **third quadrant**, one to the left and below the origin, both x and y are negative. This is denoted by $(-a,-b)$ in Fig 2.

    d. In the **fourth quadrant**, one to the right and below the origin, x is positive, while y is negative.This is denoted by $(a,-b)$ in Fig 2.

10. When the axes are not perpendicular to each other, they are said to be oblique Axes. The Greek letter "omega"- ω is used to denote the angle between their two positive directions OX and OY.

11. It is however more convenient to take the axes OX and OY at right angles. They are said to be Rectangular axes. This system of co-ordinates is called Cartesian rectangular co-ordinate system.

12. Much of the material that follows uses the Cartesian rectangular coordinate system.

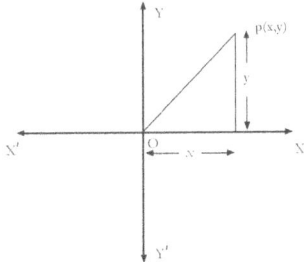

Fig. 3: Cartesian Rectangular Coordinate System

13. Let us determine distance between two points P1 (x_1, y_1) and P2 (x_2, y_2).

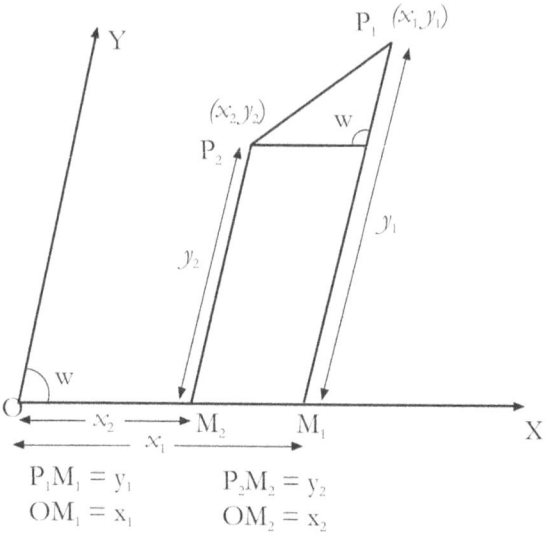

$P_1M_1 = y_1$ $P_2M_2 = y_2$
$OM_1 = x_1$ $OM_2 = x_2$

Fig. 4

Let P_1 and P_2 be the two given points, and let their co-ordinates be respectively (x_1, y_1) and (x_2, y_2)

Draw P_1M_1 and P_2M_2 parallel to OY, to meet OX in M_1 and M_2.

Draw P_2R parallel to OX to meet M_1P_1 in R.

Then,

$P_2R = M_2M_1 = OM_1 - OM_2$

$= x_1 - x_2$

$RP_1 = M_1P_1 - RM_1$

$= M_1P_1 - P_2M_2 \ (\because P_2M_2 = RM_1)$

$= y_1 - y_2$

And

$$\lfloor P_2RP_1 = \lfloor OM_1P_1 = 180° - \lfloor P_1M_1X$$

$$= 180° - \omega.$$

We therefore have

$$P_1P_2^2 = P_2R^2 + P_1R^2 - 2P_2R \cdot P_1R \cos P_2RP_1$$

$$= (x_1 - x_2)^2 + (y_1 - y_2)^2 - 2(x_1 - x_2)(y_1 - y_2)\cos(180° - \omega)$$

$$= (x_1 - x_2)^2 + (y_1 - y_2)^2 - 2(x_1 - x_2)(y_1 - y_2)(-\cos\omega)$$

$$P_1P_2^2 = (x_1 - x_2)^2 + (y_1 - y_2)^2 + 2(x_1 - x_2)(y_1 - y_2)\cos\omega$$

If the axes be, as in general the case, at right angle, we have $\omega = 90°$ and hence $\cos\omega = \cos 90° = 0$.

The formula becomes

$$P_1P_2^2 = (x_1 - x_2)^2 + (y_1 - y_2)^2$$

Or $P_1P = \sqrt{(x_1 - x_2)^2 + (y_1 - y_2)^2}$

So that in rectangular co-ordinates the distance between the two points (x_1, y_1) and (x_2, y_2) is

$$\sqrt{(x_1 - x_2)^2 + (y_1 - y_2)^2}$$

14. Corollary: The distance of the point (x_1, y_1) from the origin is

$$\sqrt{(x_1 - 0)^2 + (y_1 - 0)^2} = \sqrt{x_1^2 + y_1^2}$$

15. To find the co-ordinates of the point which divides in a given ratio $(m_1 : m_2)$ the line joining two given points (x_1, y_1) and (x_2, y_2).

Let P_1 be the point (x_1, y_1), P_2 be the point (x_2, y_2) and P be the required point, so that we have

$$P_1P : PP_2 = m_1 : m_2$$

$$\frac{P_1P}{PP_2} = \frac{m_1}{m_2}$$

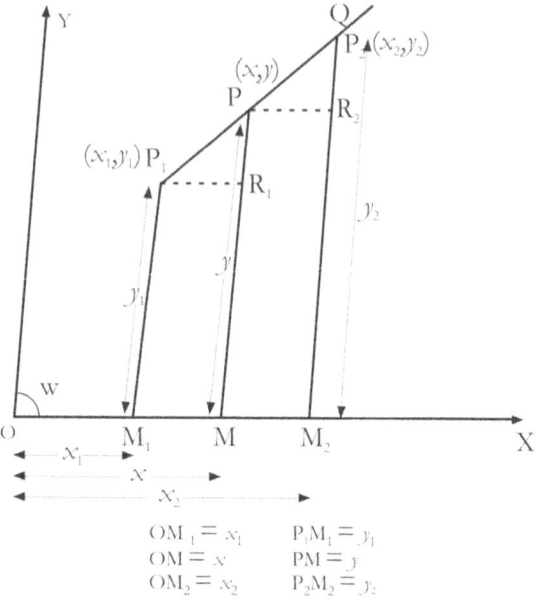

Fig 5

Let P be the point (x, y), such that if P_1M_1, PM and P_2M_2 be drawn parallel to the axis of y to meet the axis of x in M_1, M and M_2, we have

$$OM_1 = x, \quad M_1P_1 = y_1, \quad OM = x, \quad MP = y, \quad OM_2 = x_2,$$
$$M_2P_2 = y_2.$$

Draw P_1R_1 and PR_2, parallel to OX, to meet MP and M_2P_2 in R_1 and R_2 respectively. Then,

$$P_1R_1 = M_1M = OM - OM_1 = x - x_1$$
$$PR_2 = MM_2 = OM_2 - OM = x_2 - x$$
$$R_1P = PM - MR_1 = PM - P_1M_1 = y - y_1$$
$$R_2P_2 = P_2M_2 - R_2M_2 = P_2M_2 - MP = y_2 - y$$

From the similar triangles P_1R_1P and PR_2P_2, we have

$$\frac{m_1}{m_2} = \frac{P_1P}{PP_2} = \frac{P_1R_1}{PR_2} = \frac{PR_1}{P_2R_2} \quad (1)$$

Consider $\dfrac{m_1}{m_2} = \dfrac{P_1R_1}{PR_2}$

i.e. $\dfrac{m_1}{m_2} = \dfrac{x - x_1}{x_2 - x}$

$m_1(x_2 - x) = m_2(x - x_1)$

$m_1x_2 - m_1x = m_2x - m_2x_1$

Or $m_2x + m_1x = m_1x_2 + m_2x_1$

$x(m_1 + m_2) = m_1x_2 + m_2x_1$

Or $x = \dfrac{m_1x_2 + m_2x_1}{m_1 + m_2}$

Again form (1)

$$\frac{m_1}{m_2} = \frac{PR_1}{P_2R_2}$$

$$\frac{m_1}{m_2} = \frac{y - y_1}{y_2 - y}$$

$m_1(y_2 - y) = m_2(y - y_1)$

$m_1y_2 - m_1y = m_2y - m_2y_1$

$y(m_1 + m_2) = m_1y_2 + m_2y_1$

$$y = \frac{m_1y_2 + m_2y_1}{m_1 + m_2}$$

The coordinates of the point $P(x, y)$ which divides P_1P_2 internally in the ratio $m_1 : m_2$ are

$$P(x, y) = \left(\frac{m_1x_2 + m_2x_1}{m_1 + m_2}, \frac{m_1y_2 + m_2y_1}{m_1 + m_2} \right)$$

If the point Q divide the line P_1P_2 externally in the same ratio, $m_1 : m_2$

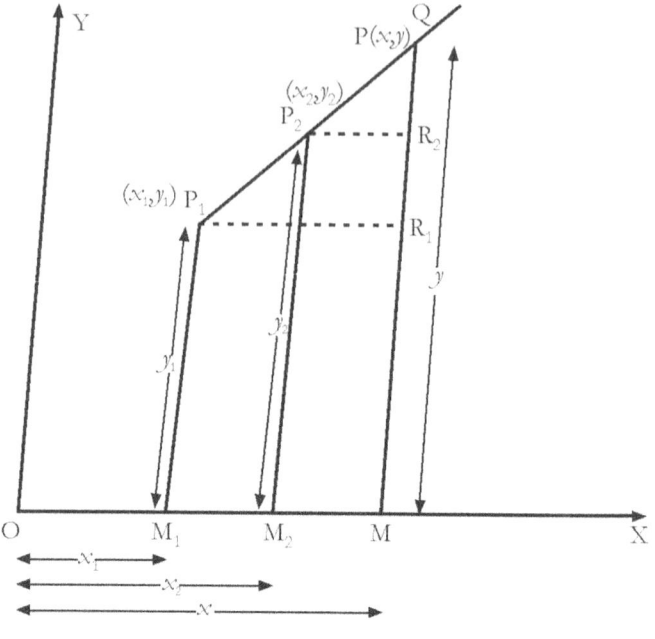

Fig. 6

Let P_1 be the point (x_1, y_1), P_2 be the point (x_2, y_2) and Q be the required point, so that we have

$$P_1Q : QP_2 = m_1 : m_2$$

Let $Q(x, y)$ be the point, which cuts P_1P_2 externally in the raio $m_1 : m_2$

If P_1M_1, P_2M_2 and QM be drawn parallel to the axis of y to meet the axis of x in M_1, M_2 and M.

We have

$$OM_1 = x_1, \; P_1M_1 = y_1, \; OM_2 = x_2, \; P_2M_2 = y_2$$

And

$$OM = x, \; QM = y.$$

Draw P_1R_1 and P_2R_2 parallel to OX, to meet QM in R_1 and R_2 respectively.

Then

$$P_1R_1 = M_1M = OM - OM_1 = x - x_1$$

$$P_2 R_2 = M_2 M = OM - OM_2 = x - x_2 .$$

$$QR_1 = QM - R_1 M = QM - P_1 M_1 = y - y_1$$

$$QR_2 = QM - R_2 M = QM - P_2 M_2 = y - y_2 .$$

From the similar triangles $P_1 R_1 P_2$ and $P_2 R_2 Q$, we have

$$\frac{m_1}{m_2} = \frac{P_1 Q}{P_2 Q} = \frac{P_1 R_1}{P_2 R_2} = \frac{QR_1}{QR_2}$$

Consider $\dfrac{m_1}{m_2} = \dfrac{P_1 R_1}{P_2 R_2}$

Or $\dfrac{m_1}{m_2} = \dfrac{x - x_1}{x - x_2}$

Or $m_1 (x - x_2) = m_2 (x - x_1)$

$$m_1 x - m_1 x_2 = m_2 x - m_2 x_1$$

$$x(m_1 - m_2) = m_1 x_2 - m_2 x_1$$

$$\therefore x = \frac{m_1 x_2 - m_2 x_1}{m_1 - m_2}$$

Again $\dfrac{m_1}{m_2} = \dfrac{QR_1}{QR_2}$

$$\frac{m_1}{m_2} = \frac{y - y_1}{y - y_2}$$

$$m_1 (y - y_2) = m_2 (y - y_1)$$

$$m_1 y - m_1 y_2 = m_2 y - m_2 y_1$$

$$y(m_1 - m_2) = m_1 y_2 - m_2 y_1$$

$$y = \frac{m_1 y_2 - m_2 y_1}{m_1 - m_2}$$

If Q divides $P_1 P_2$ externally in the ratio $m_1 m_2$ then its co-ordinates is given by

$$Q(x, y) = \left(\frac{m_1 x_2 - m_2 x_1}{m_1 - m_2}, \frac{m_1 y_2 - m_2 y_1}{m_1 - m_2} \right)$$

Above formula is obtained from internal division formula by just replacing $m_1 = m_1$ & $m_2 = -m_2$

$$P(x, y) = \left(\frac{m_1 x_2 + m_2 x_1}{m_1 + m_2}, \frac{m_1 y_2 + m_2 y_1}{m_1 + m_2} \right)$$

$m_1 = m_1$, $m_2 = -m_2$ gives

$$Q(x, y) = \left(\frac{m_1 x_2 + (-m_2) x_1}{m_1 + (-m_2)}, \frac{m_1 y_2 + (-m_2) y_1}{m_1 + (-m_2)} \right)$$

$$= \left(\frac{m_1 x_2 - m_2 x_1}{m_1 - m_2}, \frac{m_1 y_2 - m_2 y_1}{m_1 - m_2} \right)$$

Corollary: The co-ordinate of the middle point of the line joining (x_1, y_1) to $(x_2 y_2)$ is $\left(\frac{x_1 + x_2}{2}, \frac{y_1 + y_2}{2} \right)$

We know that the midpoint divides any line in 1:1 ratio. Thus replacing $m_1 = 1$, $m_2 = 1$ in internal division formula we get

mid point $P(x, y) = \left(\frac{x_1 + x_2}{2}, \frac{y_1 + y_2}{2} \right)$

16. Derive a formula for determining the area of the triangle, the co-ordinates of whose angular points given, the axes being rectangular.

Let ABC be the triangle and let the co-ordinates of its angular points, A, B and C be (x_1, y_1) (x_2, y_2) & (x_3, y_3)

Draw AL, BM and CN perpendicular to the axis of x and let Δ dente the required area.

Then,

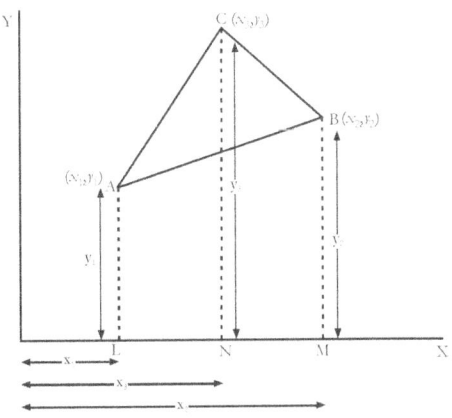

Fig. 7

Δ = trapezium ALNC + trapezium CNMB − trapezium ALMB

$$=\frac{1}{2}LN(AL+NC)+\frac{1}{2}NM(NC+MB)-\frac{1}{2}LM(LA+MB)$$

$LN = ON - OL = x_3 - x_1,\ LA = y_1, NC = y_3$

$NM = OM - ON = x_2 - x_3,\ MB = y_2$

$LM = OM - OL = x_2 - x_1$

$$\therefore \Delta=\frac{1}{2}(x_3-x_1)(y_1+y_3)+\frac{1}{2}(x_2-x_3)(y_3+y_2)$$

$$-\frac{1}{2}(x_2-x_1)(y_1+y_2)$$

$$=\frac{1}{2}\{x_3y_1+\cancel{x_3y_3}-\cancel{x_1y_1}-x_1y_3+x_2y_3+\cancel{x_2y_2}-\cancel{x_3y_3}$$

$$-x_3y_2-x_2y_1-\cancel{x_2y_2}+\cancel{x_1y_1}+x_1y_2\}$$

$$=\frac{1}{2}\{x_1(y_2-y_3)+x_2(y_3-y_1)+x_3(y_1-y_2)\}$$

If we use the determinant notation this may be written as

$$\Delta=\frac{1}{2}\begin{vmatrix} x_1 & y_1 & 1 \\ x_2 & y_2 & 1 \\ x_3 & y_3 & 1 \end{vmatrix}$$

17. Corollary: The area of the triangle whose vertices are the origin $(0,0)$ and the (x_1,y_1) & (x_2,y_2) is $\Delta=\frac{1}{2}(x_1 y_2 - x_2 y_1)$

If the axes be oblique, the perpendicular AL, BM and CM are not equal to the ordinates y_1, y_2 and y_3 but are equal respectively to $y_1 \sin \omega, y_2 \sin \omega,$ and $y_3 \sin \omega$ .Then the area of the triangle in this case becomes,

$$\Delta=\frac{1}{2}\sin \omega \left\{ x_1 y_2 - x_2 y_1 + x_2 y_3 - x_3 y_2 + x_3 y_1 - x_1 y_3 \right\}$$

$$\Delta=\frac{1}{2}\sin \omega \begin{vmatrix} x_1 & y_1 & 1 \\ x_2 & y_2 & 1 \\ x_2 & y_3 & 1 \end{vmatrix}$$

18. Derive a formula for the area of a quadrilateral the co-ordinates of whose angular points are given.

Let the angular points of the quadrilateral, taken in order, be A, B, C and D and let their co-ordinates be respectively (x_1, y_1) (x_2, y_2) (x_3, y_3) & (x_4, y_4)

Draw AL, BM, CN& DR perpendicular to the axis of x

Then the area of the quandrilateral

= trapezium ALRD + trapezium DRNC + trapezium CNMB - trapezium ALMB

$$=\frac{1}{2}LR(LA+RD)+\frac{1}{2}RN(RD+NC)+\frac{1}{2}NM(NC+MB)$$

$$-\frac{1}{2}LM(LA+MB)$$

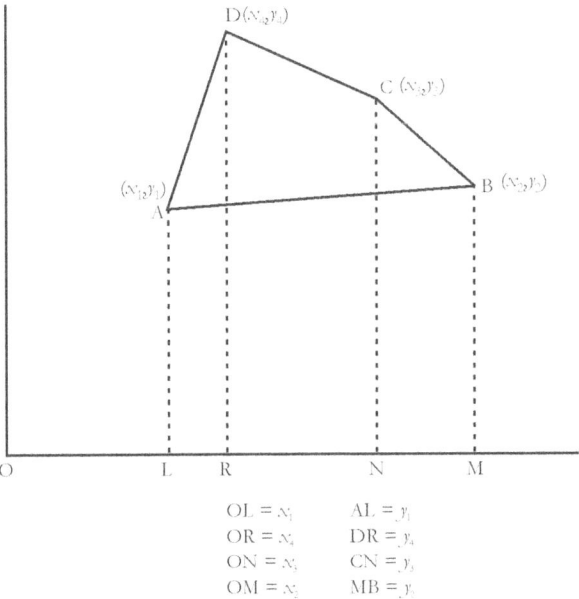

OL = x_1 AL = y_1
OR = x_4 DR = y_4
ON = x_3 CN = y_3
OM = x_2 MB = y_2

Fig. 8

We have

$$LR = OR - OL = x_4 - x_1$$

$$RN = ON - OR = x_3 - x_4$$

$$NM = OM - ON = x_2 - x_3$$

$$LM = OM - OL = x_2 - x_1$$

$$\therefore \Delta = \frac{1}{2}\{(x_4 - x_1)(y_1 + y_4) + (x_3 - x_4)(y_3 + y_4) + (x_2 - x_3)$$

$$(y_3 + y_2) - (x_2 - x_1)(y_1 + y_2)\}$$

$$\Delta = \frac{1}{2}\{(x_1 y_2 - x_2 y_1) + (x_2 y_3 - x_3 y_2) + (x_3 y_4 - x_4 y_3)$$

$$+ (x_4 y_1 - x_1 y_4)\}$$

In a similar manner it may be shown that the area of a polygon of n sides the co-ordinates of whose angular points, taken in order are

$(x_1, y_1)(x_2, y_2), \cdots (x_n, y_n)$ is

$$\Delta = \frac{1}{2}\left\{(x_1 y_2 - x_2 y_1) + (x_2 y_3 - x_3 y_2) + \cdots + (x_n y_1 - x_1 y_2)\right\}$$

19. Polar co-ordinates: Suppose, O be a fixed point called the origin or pole, and OX a fixed line called the initial line.

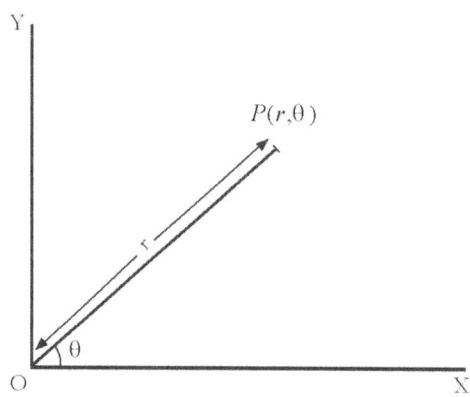

Fig. 9

a. Take any other point P in the plane of the paper and join OP. The position of P is clearly known when the angle XOP and the length OP are given.

b. If the vectorial angle θ and the radius vector be r, the position of P is denoted by the symbol (r, θ), where r and θ are called polar co-ordinates of P.

c. The radius vector is positive if it be measured from the origin O along the line bounding the vectorial angle; if measured in the opposite direction it is negative.

20. To find the area of a triangle the co-ordinates of whose angular points are given.

Let ABC be the triangle and let $(r_1, \theta_1), (r_2, \theta_2)$ and (r_3, θ_3) be the polar coordinates of its angular points. We have

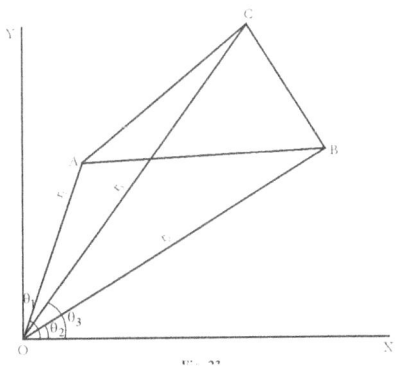

Fig 10

$$\triangle ABC = \triangle OBC + \triangle OCA - \triangle OBA \quad (1)$$

Now

$$\triangle OBC = \frac{1}{2} OB.OC \sin \lfloor BOC$$

$$= \frac{1}{2} r_2 r_3 \sin(\theta_3 - \theta_2)$$

$$\because \lfloor BOC = \lfloor COX - \lfloor BOX$$

$$= \theta_3 - \theta_2$$

So

$$\triangle OCA = \frac{1}{2} OC.OA \sin \lfloor COA$$

$$= \frac{1}{2} r_3 r_2 \sin(\theta_1 - \theta_3)$$

$$\left[\because \lfloor COA = \lfloor AOX - \lfloor COX = \theta_1 - \theta_3 \right]$$

And $\triangle OAB = \frac{1}{2} OA.OB \sin AOB$

$$= \frac{1}{2} r_1 r_2 \sin(\theta_1 - \theta_2) \left[\because \lfloor AOB = \lfloor AOX - \lfloor BOX = \theta_1 - \theta_2 \right]$$

$$= \frac{-1}{2} r_1 r_2 \sin(\theta_2 - \theta_1)$$

Hence (1) gives

$$\triangle ABC = \frac{1}{2}\left[r_2 r_3 \sin(\theta_3 - \theta_2) + r_3 r_1 \sin(\theta_1 - \theta_3) + r_1 r_2 \sin(\theta_2 - \theta_1)\right]$$

21. To change from Cartesian co-ordinates to polar co-ordinates, and conversely

    Let P be any point whose Cartesian co-ordinates, referred to rectangular axes, are x and y, and whose polar co-ordinates, referred to O as pole and OX as initial line are (r, θ)

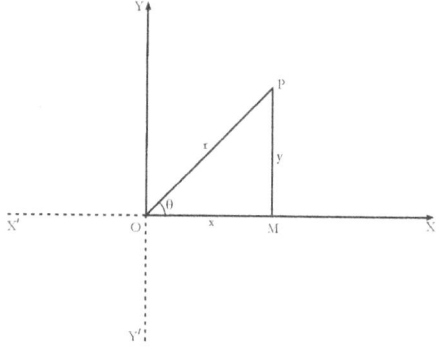

Fig. 11

Draw PM perpendicular to OX

So that we have

$$OM = x, \quad MP = y \quad \underline{|MOP} = \theta$$

And $OP = r$

From the triangle MOP, we have

$$x = OM = OP\cos\underline{|MOP} \quad \left| \begin{array}{l} \because \cos\underline{|MOP} = \dfrac{OM}{OP} \\ OP\cos\underline{|MOP} = OM \end{array} \right.$$

$$= r\cos\theta \quad (1)$$

$$y = MP = OP \sin \lfloor MOP \quad \left| \begin{matrix} \because \sin \lfloor MOP = \dfrac{PM}{OP} \\ PM = OP \sin \lfloor MOP \end{matrix} \right.$$

$$= r \sin \theta \quad (2)$$

$$r = OP = \sqrt{OM^2 + MP^2}$$

$$= \sqrt{x^2 + y^2} \quad (3)$$

And $\tan \theta = \dfrac{MP}{OM} = \dfrac{y}{x}$ (4)

Equations (1) and (2) express the Cartesian co-ordinates in terms of the polar co-ordinates.

Equations (3) & (4) express the polar in terms of Cartesian co-ordinates.

The same relations will be found to hold if P be in any other of the quadrants into which the plane is divided by XOX' and YOY'.

Solved Problems

Example 1:- Determine the position of the points

    1. $(2, -1)$

    2. $(-3, 2)$

    3. $(-2, -3)$

    4. $(2, -1)$

1.   We measure a distance 2 along OX and then a distance 1 parallel to OY'. We get the point $(2, -1)$.

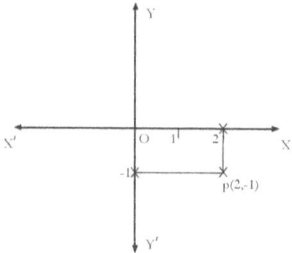

Fig. 12

2. $(-3, 2)$: We measure a distance 3 along OX' and then measure a distance 2 parallel to OY. We can now plot $(-3, 2)$

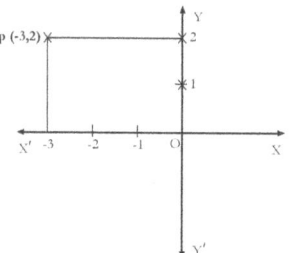

Fig. 13

3. $(-2, -3)$: We measure 2 along OX' and then 3 parallel to OY'. We can now plot the point $(-2, -3)$.

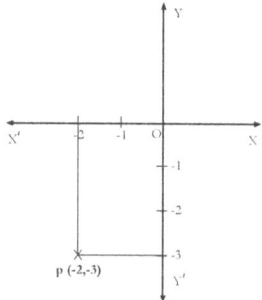

Fig. 14

Example 2:   In any triangle ABC, if D is the midpoint of BC, prove that $AB^2 + AC^2 = 2(AD^2 + DC^2)$.

Take B as origin, BC as the axis of x and a line through B perpendicular to BC as the axis of y

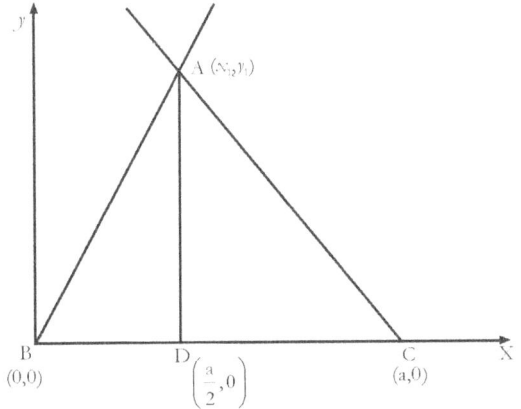

Fig. 15

Let $BC = a$, so that C is the point $(a,0)$, and let A be the point (x_1, y_1).

Then D is the point $\left(\dfrac{0+a}{2}, \dfrac{0+0}{2}\right) = \left(\dfrac{a}{2}, 0\right)$

Hence $AD^2 = \left(x_1 - \dfrac{a}{2}\right)^2 + y_1^2$ and

$DC^2 = \left(a - \dfrac{a}{2}\right)^2 + (0-0)^2 = \left(\dfrac{a}{2}\right)^2$

$2(AD^2 + DC^2) = 2\left[\left(x_1 - \dfrac{a}{2}\right)^2 + y_1^2 + \left(\dfrac{a}{2}\right)^2\right]$

$= 2\left[x_1^2 + \dfrac{a^2}{4} - x_1 a + y_1^2 + \dfrac{a^2}{4}\right]$

$= 2x_1^2 + 2y_1^2 - 2ax_1 + a^2$

$AC^2 = (x_1 - a)^2 + y_1^2$

$$AB^2 = -x_1^2 + y_1^2$$

Therefore $AB^2 + AC^2 = 2x_1^2 + 2y_1^2 - 2ax_1 + a^2$

Hence $AB^2 + AC^2 = 2(AD^2 + DC^2)$

Example 3:   ABC is a triangle and D,E and F are the middle points of the sides BC, CA and AB; prove that the point which divides AD internally in the ratio 2:1 also divides the lines BE and CF in the same ratio.

Let the co-ordinates of the vertice A,B,C be (x_1, y_1) (x_2, y_2) and (x_3, y_3) respectively.

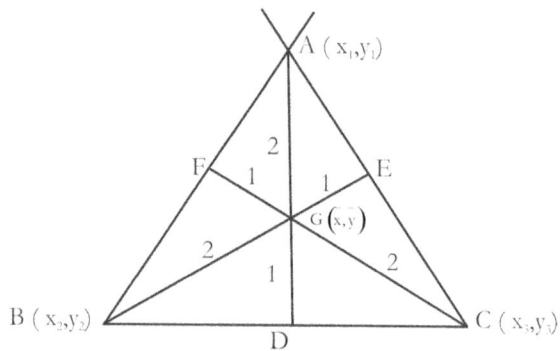

Fig. 16

The co-ordinates of D are therefore $= \left(\dfrac{x_2 + x_3}{2}, \dfrac{y_2 + y_3}{2} \right)$

Let G be the point that divides AD internally in the ratio 2:1 and let its co-ordinates be $\bar{x}$ and $\bar{y}$.

We know that general formula for internal division

$$G(\bar{x}, \bar{y}) = \left(\frac{m_1 x_2 + m_2 x_1}{m_1 + m_2}, \frac{m_1 y_2 + m_2 y_1}{m_1 + m_2} \right)$$

$m_1 = 2$

$m_2 = 1$.

$$= \left[\frac{\cancel{2}\left(\dfrac{x_2 + x_3}{\cancel{2}}\right) + |x| }{2 + 1} , \frac{\cancel{2}\left(\dfrac{y_2 + y_3}{\cancel{2}}\right) + 1 \cdot y_1}{2 + 1} \right]$$

$$G(\overline{x}, \overline{y}) = \left(\frac{x_1 + x_2 + x_3}{3}, \frac{y_1 + y_2 + y_3}{3} \right)$$

In the same manner we could show that there are the co-ordinates of the points that divide BE and CF in the ratio 2:1.

Since, a point whose co-ordinates are

$$\left(\frac{x_1 + x_2 + x_3}{3}, \frac{y_1 + y_2 + y_3}{3} \right)$$

lies on each of the lines AD, BE and CF, it follows that these three lines meet in a point. *This point is called the centroid of the triangle.*

Example 3: To prove that the area of a trapezium, i.e. a quadrilateral having two sides parallel is one half the sums of the two parallel sides multiplied by the perpendicular distance between them.

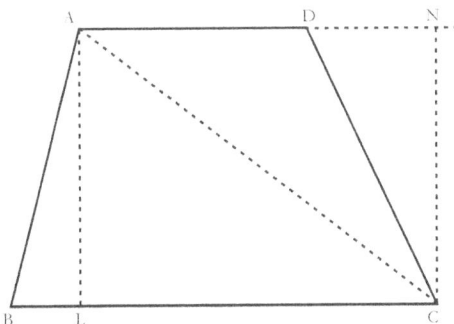

Fig 17

Let ABCD be the trapezium having the sides AD & BC parallel. Join AC and draw AL perpendicular to BC and CN perpendicular to AD, produced if necessary. Since, the area of a triangle is one half the product of any side and the perpendicular drawn from the opposite angle, we have:

Area ABCD = area of $\triangle ABC$ + area of $\triangle ACD$.

$$= \frac{1}{2} BC \times AL + \frac{1}{2} \times AD \times CN$$

$$= \frac{1}{2}(BC + AD) \times AL \; (\because AL = CN)$$

Example 6: Construct the positions of the points

(i) $(2, 30°)$

(ii) $(3, 150°)$,

(iii) $(-2, 45°)$

(iv) $(-3, 330°)$

(v) $(3, -210°)$

(vi) $(-3, -30°)$

i. $(2, 30°)$

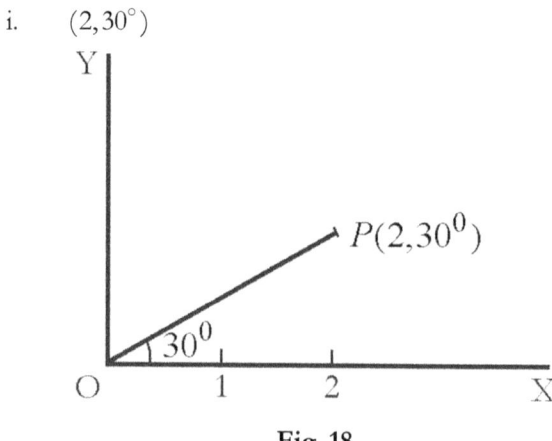

Fig. 18

Let the radius vector revolve from OX through an angle of $30°$, and then mark off along it a distance equal to two units of length.

We thus obtain $P_1(2,30°)$.

ii. $(3,150°)$

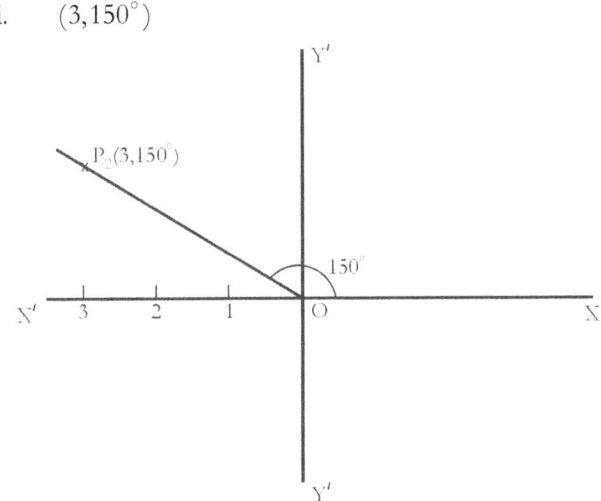

Fig. 19

iii. $(-2,45°)$

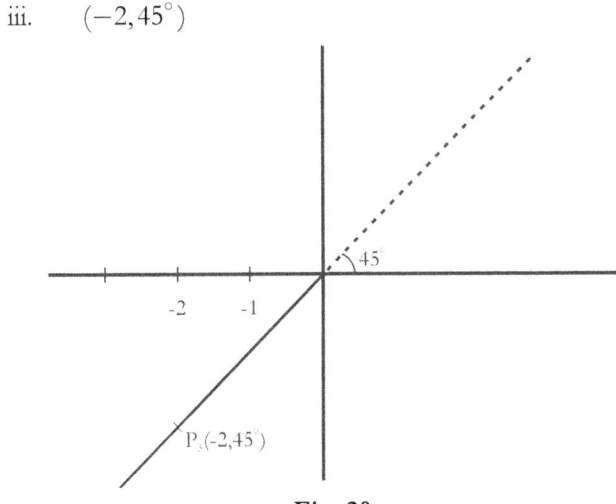

Fig. 20

Let the radius vector revolve from OX through 45° into the position OL. We have now to measure along OL a distance −2

i.e. We have to measure a distance 2 not along OL but in the opposite direction producing LO to P_3 so that OP_3 is 2 units of length, we have the required point P_3.

iv. $(-3, 330°)$

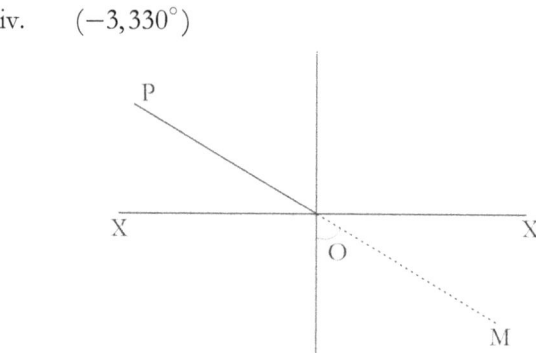

Fig. 21

To get this point, we let the radius vector rotate from OX through 330° into the position OM and measure on it a distance −3 i.e. 3 in the directions MO produced. We thus, have the point P_2, which is the same as the point given by (ii)

v. $(3, -210°)$

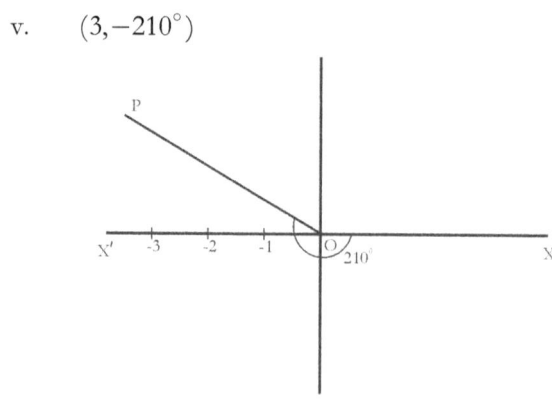

Fig. 21

If the radius vector rotate through $-210°$ it will be in the position OP_2, and the point required is P_2.

vi. $(-3, -30°)$

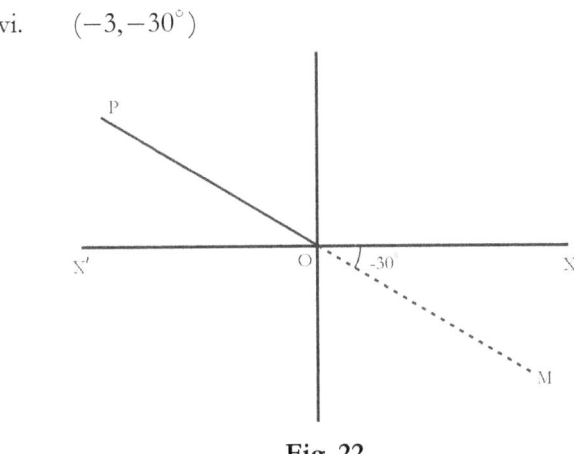

Fig. 22

The radius vector, after rotating through $-30°$ is in the position OM. We then measure -3 along it i.e. 3 in the direction MO produced and once more arrive at the point B.

In general it will be found that the same point is given by each of the polar co-ordinates

$$(r, \theta)\,(-r, 180° + \theta),(r, -(360 - \theta))\text{ and }(-r, -(180 - \theta))$$

Or expressing the angles in radians, by each of the co-ordinate

$$(r, \theta)(-r, \pi + \theta)(r, -(2\pi - \theta))\text{ and }\{-r, -(\pi - \theta)\}$$

It is also clear that adding $360°$ to the vectorial angle does not alter the final position of the revolving line, so that (r, θ) is always the same point as $(r, \theta + n \cdot 360°)$, where n is an integer.

So adding $180°$ or any odd multiple of $180°$ to the vectorial angle and changing the sign of the radius vector gives the same point as before. Thus the point

$$\left[-r, \theta + (2n + 1)180°\right]$$

Is the same point as $[-r, \theta + 180°]$ i.e. is the point (r, θ).

Example 9: Determine Cartesian equations for the following po-lar coordinates (1) $r = a \sin \theta$ **(2)** $r^{1/2} = a^{1/2} \cos \dfrac{\theta}{2}$.

1. We are given:

$$r = a \sin \theta$$

$$r^2 = ar \sin \theta$$

We know that $r^2 = x^2 + y^2$ and $y = r \sin \theta$

$$\therefore x^2 + y^2 = ay$$

2. $\quad r^{1/2} = a^{1/2} \cos \dfrac{\theta}{2}$

Squaring both sides, we get

$$r = a \cos^2 \frac{\theta}{2}$$

$$= a \left(\frac{1 + \cos \theta}{2} \right)$$

$$2r = a(1 + \cos \theta)$$

$$2r^2 = ar(1 + \cos \theta)$$

$$2(x^2 + y^2) = a\sqrt{x^2 + y^2} + ax \quad \begin{bmatrix} \because r^2 = x^2 + y^2 \\ r \cos \theta = x \end{bmatrix}$$

Or $2x^2 + 2y^2 - ax = a\sqrt{x^2 + y^2}$

Squaring both sides

$$(2x^2 + 2y^2 - ax)^2 = a^2(x^2 + y^2)$$

3. Locus

1. When a point moves so as always satisfy a given condition, or conditions, the path it traces out is called its **Locus** under these conditions.

2. For example, let any point $P(x, y)$ moves in a plane such that whose distance from a fixed point O is always constant. This locus is the circumference of the circle, where O is the centre and constant distance is its radius. Clearly,

    $$OP_1' = OP_2' = OP_3' = OP_5' = \text{constant} = \text{radius}$$

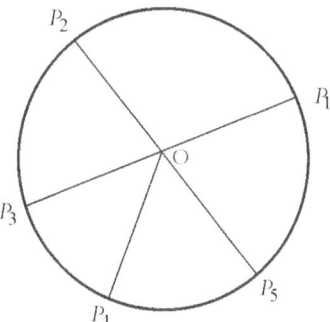

Fig 23

3. Let us consider another case. Suppose A and B be two fixed points P is to move in the plane of the paper so that its distance from A and B are to be always equal. If we bisect AB in C and through it draw a straight line 1 to AB then any point on this straight line is at equal distances from A and B.i.e.

    $$AP_1 = P_1B$$

    $$AP_2 = P_2B$$

    $$AP_3 = P_3B$$

The perpendicular bisector is the locus of P subject to the assumed condition.

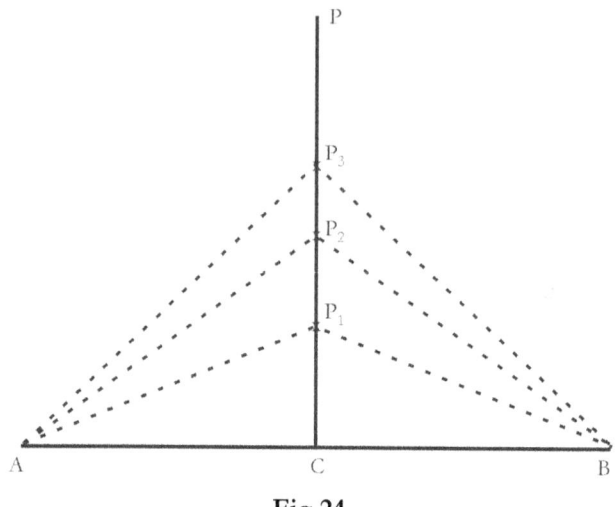

Fig 24

4. Again, if A and B are two fixed points and that the point P is to move in the plane of the paper so that the angle APB is always a right angle.If we describe a circle on AB as diameter then P may be any point on the circumference of this circle, since the angle in a semicircle is a right angle, also it could easily be shown that APB is met aright angle except when P lies on this circumference. The locus's of P under the assumed condition is therefore a circle on AB as diameter.

5. One single equation between two unknown quantities x and y, $x + y = 1$. In this case, we cannot completely determine the values of x and y.Such an equation has an infinite number of solutions.Amongst them are the following

$$
\left.\begin{array}{l} x=0 \\ y=1 \end{array}\right] \left.\begin{array}{l} x=1 \\ y=0 \end{array}\right] \left.\begin{array}{l} x=2 \\ y=-1 \end{array}\right\} \left.\begin{array}{l} x=3 \\ y=-2 \end{array}\right\} \cdots \left.\begin{array}{l} x=-1 \\ y=2 \end{array}\right] \left.\begin{array}{l} x=-2 \\ y=3 \end{array}\right\} \cdots
$$

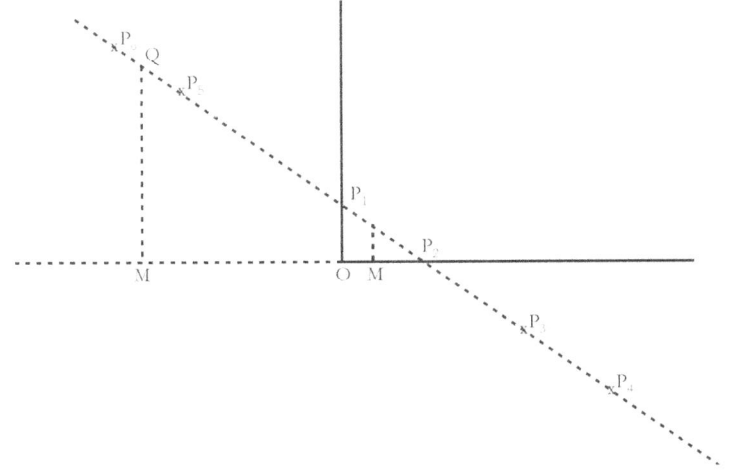

Fig 25

Let OX and OY be the axes of co-ordinates.

If we mark off a distance $OP_1(-1)$ along OY, we have a point P_1 whose co-ordinates $(0, 1)$ clearly satisfy equation (1)

If we mark off a distance $OP_2(=1)$ long OX, we have a point P_2 whose co-ordinates $(1, 0)$ satisfy equation (1)

Similarly the point $P_3(2,-1)$ and $P_4(3,-2)$ satisfy the equation (1).

Again, the co-ordinates $(-1, 2)$ of P_5 and the co-ordinates $(-2, 3)$ of P_6 satisfy equation (1).

On making the measurements carefully we should find that all the points we obtain lie on the line P_1P_2 (produced both ways)

Thus $x + y = 1$ is the equation to the straight line P_1P_2.

6. Let us now turn our attention to another equation. $x^2 + y^2 = 4$. Amongst an infinite number of solutions of this equation are the following.

$$\left.\begin{array}{l} x = 2 \\ y = 0 \end{array}\right] \left.\begin{array}{l} x = \sqrt{3} \\ y = 1 \end{array}\right] \left.\begin{array}{l} x = \sqrt{2} \\ y = \sqrt{2} \end{array}\right] \left.\begin{array}{l} x = 1 \\ y = \sqrt{3} \end{array}\right]$$

$$\left. \begin{matrix} x=0 \\ y=2 \end{matrix} \right\} \quad \left. \begin{matrix} x=-1 \\ y=\sqrt{3} \end{matrix} \right\} \quad \left. \begin{matrix} x=-\sqrt{2} \\ y=\sqrt{2} \end{matrix} \right\} \quad \left. \begin{matrix} x=-\sqrt{3} \\ y=1 \end{matrix} \right\}$$

$$\left. \begin{matrix} x=-2 \\ y=0 \end{matrix} \right\} \quad \left. \begin{matrix} x=-\sqrt{3} \\ y=-1 \end{matrix} \right\} \quad \left. \begin{matrix} x=-\sqrt{2} \\ y=-\sqrt{2} \end{matrix} \right\} \quad \left. \begin{matrix} x=-1 \\ y=-\sqrt{3} \end{matrix} \right\}$$

$$\left. \begin{matrix} x=0 \\ y=-2 \end{matrix} \right\} \quad \left. \begin{matrix} x=1 \\ y=-\sqrt{3} \end{matrix} \right\} \quad \left. \begin{matrix} x=\sqrt{2} \\ y=-\sqrt{2} \end{matrix} \right\} \quad \left. \begin{matrix} x=\sqrt{3} \\ y=-1 \end{matrix} \right\}$$

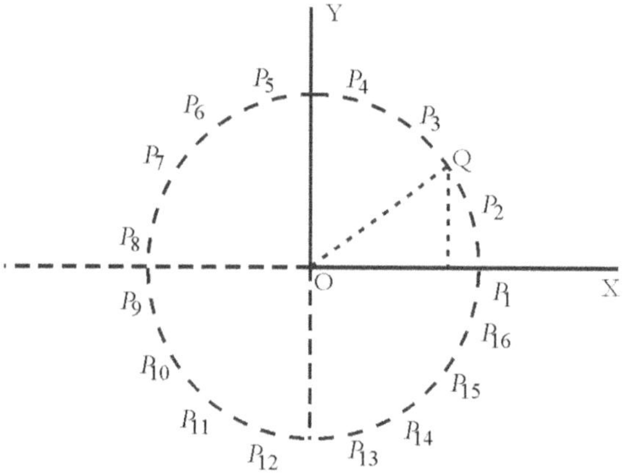

Fig 26

All these point are respectively represented by the points $P_1, P_2, \cdots P_{16}$ and they will all be found to lie on the dotted circle whose centre is O and radius is 2.

Also, if we take any point Q on this circle and its ordinate QM, it follows, since $OM^2 + MQ^2 = OQ^2 = 4$ that the x and y of the point Q satisfy (1).

The dotted circle therefore passes through all the points whose co-ordinates satisfy (1)

7. **Equation to a curve**: The equation to a curve is the relation which exists between the coordinates of any point on the curve, and which holds for no other points except those lying on the curve.

8. Conversely, to every equation between x and y it will be found that there is, in general, a definite geometrical locus.

    a. The equation is $x + y = 1$, and the definite path or locus is the straight line.

    b. The equation is $x^2 + y^2 = 4$, and the definite path or locus is the circle.

    c. Again the equation $y = 1$ states that the moving point is such that its ordinate is always unity that it is always at a distance 1 from the axis of x. The definite path or locus is therefore a straight line parallel to OX end at a distance unity from it.

Solved Examples

Example 1: A point moves so that the algebraic sum of its distances from two given perpendicular axes is equal to a constant quantity a, find the equation to its locus.

> Take the two straight lines as the axes of coordinates. Let (x, y) be any point satisfying the given condition, we then have $x + y = a$. This being the relation connecting the coordinates of any point on the locus is the equation to the locus.

Example 2: The sum of the squares of the distances of a moving point from the two fixed points $(a, 0)$ and $(-a, 0)$ is equal to a constant quantity $2c^2$. Find the equation to its locus.

> Let (x, y) be any position of the moving point. The condition given in the question can be represented as:
>
> $$\{(x-a)^2 + y^2\} + \{(x+a)^2 + y^2\} = 2c^2$$
>
> i.e. $x^2 + a^2 - 2\!\!\!\!\diagup\!\!xa + y^2 + x^2 + a^2 + 2\!\!\!\!\diagup\!\!xa + y^2 = 2c^2$
>
> $$2x^2 + 2y^2 + 2a^2 = 2c^2$$

Or $x^2 + y^2 = c^2 - a^2$

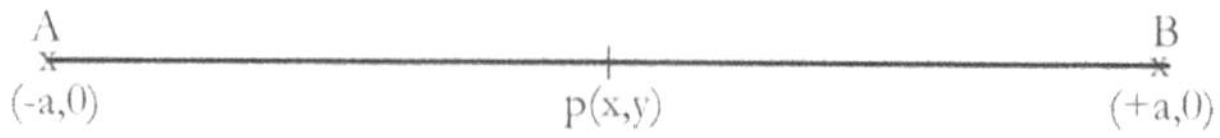

Fig 27

This being the relation between the co-ordinates of any, and every point that satisfies the given condition

This equation tells us that the square of the distance of the point (x, y) from the origin is constant and equal to $c^2 - a^2$, and therefore the locus of the point is a circle whose centre is the origin.

Example 3: A point moves so that its distance from the point $(-1, 0)$ is always three times its distance from the point $(0, 2)$

Let (x, y) be any point which satisfies the given condition. We then have

$$\sqrt{(x+1)^2 + (y-0)^2} = 3\sqrt{(x-0)^2 + (y-2)^2}$$

So that, on squaring

$$(x+1)^2 + y^2 = 9[x^2 + (y-2)^2]$$

$$x^2 + 2x + 1 + y^2 = 9x^2 + 9y^2 + 36 - 36y$$

$$8(x^2 + y^2) - 2x - 36y + 35 = 0$$

This is the relation between the co-ordinates of each, and every point that satisfies the given relation. Hence this equation defines the required locus.

4. Straight Line: Rectangular Coordinates

Let us understand straightlines and representation of straightlines in Cartesian coordinate system.

1. **Let us determine the equation to a straight line which is parallel to one of the coordinate axis**

   Let CL be any line parallel to the axis of y and passing through a point C on the axis of x such that $OC = c$.

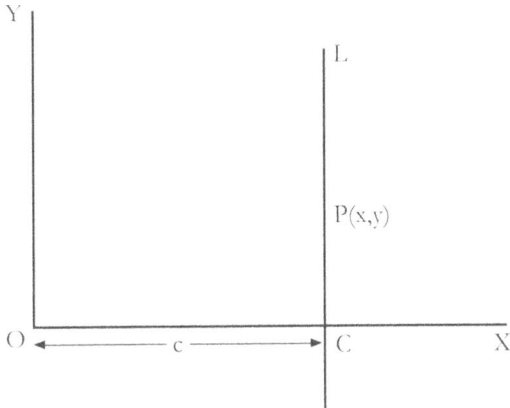

Fig. 28

   Let P be any point on this line whose co-ordinates are x and y.

   Then the abscissa of the point P is always c. So that

   $$x = c \quad (1)$$

   Similarly, the equation to a straight line parallel to the axis of x is $y = d$. [Refer Fig 29]

2. **Corollary:**As a consequence of the discussion, we can make the following observations. The equation to the axis of x is $y = 0$ .The equation to the axis of y is $x = 0$ ..

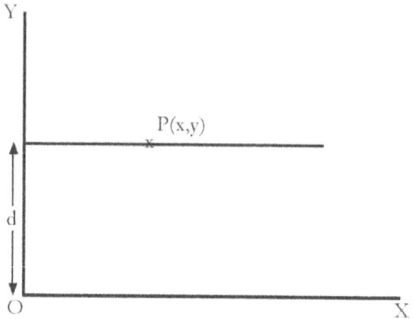

Fig. 29

3. **To find the equation to a straight line which cuts off a given intercept on the of y and is inclined at a given angle to the axis of x.**

Let the given intercept be c and let the given angle be α.

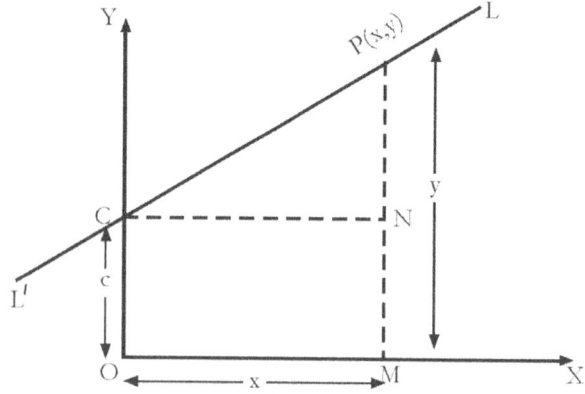

Fig. 30

Let C be a point on the axis of y such that OC is c.

Through C draw a straight line LCL' inclined at an angle $\alpha(= \tan^{-1} m)$ to the axis of x,

So that $m = \tan \alpha$

Draw $PM \perp$ to OX to meet in N a line through C. parallel to OX.

Let the co-ordinate of P be x and y, so that

$$OM = x \text{ and } MP = y \text{ from } \triangle^{le} CPN$$

$$\text{Then } MP = NP + NM \left(\tan\alpha = \frac{PN}{CN} \right)$$

$$= CN \tan\alpha + OC \text{ or} \left(PN = CN \tan\alpha \right)$$

$$y = x \tan\alpha + c \; (\because CN = OM = x)$$

$$y = mx + c \; (\because m = \tan\alpha)$$

4. **Corollary:** The equation to any straight line passing through the origin i.e. which cuts off a zero intercept from the axis of y, is found by putting $c = 0$ and hence is $y = mx$.

Example: The equation to any straight line cutting off an intercept 3 from the negative direction of the axis of y, and inclined at $120°$ to the axis of x, is

$$y = x \tan 120 + (-3) \; (\because \alpha = 120)$$

$$y = -\sqrt{3}x - 3 \; (C = -3)$$

Or $\sqrt{3}x + y + 3 = 0 \; \tan 120 = \tan(90 + 30)$

$$= -\cot 30°$$

$$= -\left(\sqrt{3} \right)$$

5. **To find the equation to the straight line which cuts off given intercepts a and b from the axes.**

Let A and B be on OX and OY respectively and be such that $OA = a$, and $OB = b$. Join AB and produce it indefinite both ways. Let P be any point (x, y) on this straight line, and draw PM perpendicular to OX.

By geometry, we have

$$\frac{OM}{OA} = \frac{PB}{AB}, \text{ and } \frac{MP}{OB} = \frac{AP}{AB}$$

$$\therefore \frac{OM}{OA} + \frac{MP}{OB} = \frac{PB}{AB} + \frac{AP}{AB} \quad \begin{bmatrix} \because OM = x \\ PM = y \end{bmatrix}$$

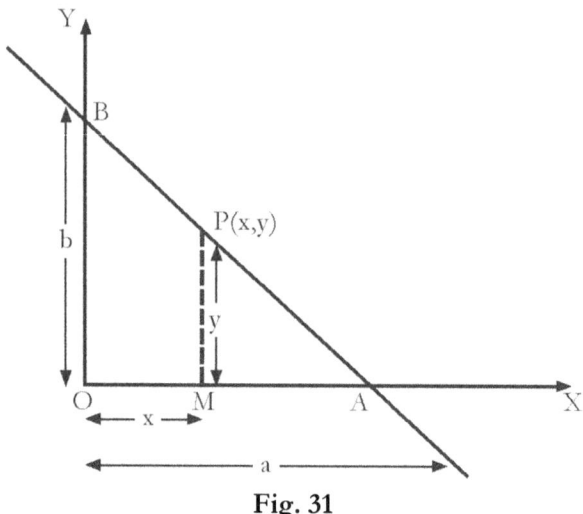

Fig. 31

$$\frac{x}{a} + \frac{y}{b} = \frac{PB + AP}{AB} \quad \begin{pmatrix} OA = a \\ OB = b \end{pmatrix}$$

$$= \frac{AB}{AB}$$

$$\therefore \frac{x}{a} + \frac{y}{b} = 1$$

6. **To find the equation to a straight line in terms of the perpendicular, let fall upon it from the origin and the angle that this perpendicular makes with the axis of x.**

   Let OR be the perpendicular from O and let its length be P

   Let α be the angle that OR makes with OX.

Let P be any point, whose co-ordinates are x and y, lying on AB; draw the ordinate PM, and also ML perpendicular to OR and PN perpendicular to ML.

Then, $OL = OM \cos \alpha$ (1)

And $LR = NP = MP \sin \lfloor NMP$

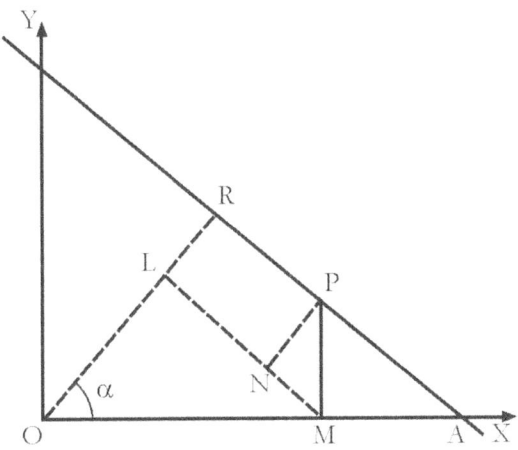

Fig. 32

But $\lfloor NMP = 90° - \lfloor NMO = \lfloor MOL = \alpha$

$$\therefore LR = MP \sin \alpha \quad (2)$$

Hence adding (1) and (2), we have

$$OM \cos \alpha + MP \sin \alpha = OL + LR = OR = P$$

$$\therefore x \cos \alpha + y \sin \alpha = P \begin{pmatrix} \because OM = x \\ MP = y \end{pmatrix}$$

This is the required equation

7. **Let us now prove that any equation of the first degree in x and y always represents a straight line.**

    $Ax + By + C = 0$ can also be written as

    $By = -Ax - C$

Or $y = \dfrac{-A}{B}x - \dfrac{C}{B}$

And this is the same as the straight line

$y = mx + C$

If $m = \dfrac{-A}{B}$, $c = \dfrac{-C}{B}$

We know that $y = mx + c$ represents equation of straight line cutting off an intercept C and inclined at an angle $\tan^{-1} m$ to the axis of x.

The equation $Ax + By + C = 0$ represents a straight line cutting off an intercept $\dfrac{-C}{B}$ from the axis of y and inclined at an angle $\tan^{-1}\left(\dfrac{-A}{B}\right)$ to the axis of x.

8. **We can reduce the general equation of the first degree equation $Ax + By + C = 0$ to the form $x\cos\alpha + y\sin\alpha = P$**

$Ax + By = C = 0$ (1)

Let P be the length of perpendicular from the origin on (1) and α the angle it makes with the axis.

∴ Equation of straight line is $x\cos\alpha + y\sin\alpha - P = 0$ (2)

(1) & (2) are same

∴ $\dfrac{\cos\alpha}{A} = \dfrac{\sin\alpha}{B} = \dfrac{-P}{C}$

i.e. $\dfrac{P}{C} = \dfrac{-\cos\alpha}{A} = \dfrac{-\sin\alpha}{B} = \dfrac{\sqrt{\cos^2\alpha + \sin^2\alpha}}{\sqrt{A^2 + B^2}} = \dfrac{1}{\sqrt{A^2 + B^2}}$

Hence $\cos\alpha = \dfrac{-A}{\sqrt{A^2 + B^2}}$ $\sin\alpha = \dfrac{-B}{\sqrt{A^2 + B^2}}$ & $P = \dfrac{C}{\sqrt{A^2 + B^2}}$

The equation (1) may therefore be reduced to the form (2) by dividing it by $\sqrt{A^2 + B^2}$ and arranging it so that the constant term is negative.

9. **To trace the straight line given by an equation of the first degree**

Let the equation be $Ax + By + C = 0$ (1)

Straight line may be traced by finding the coordinates of any two points on it.

If we put $y = 0$ in (1), we get $x = \dfrac{-C}{A}$

The point is $\left(\dfrac{-C}{A}, 0\right)$ therefore lies on it

If we put $x = 0$ in (1), we get $y = \dfrac{-C}{B}$

So the point is $\left(0, \dfrac{-C}{B}\right)$ lies on it.

10. **Straight line : A few observations**

$Ax + By + C = 0$ represents a general equation of straight line.

Case 1: If A = 0, but not B or C then $y = $ constant

$\therefore$ Straight line is $y = $ constant; a line parallel to x-axis

Case 2: If B = 0, but not A or C

$\therefore$ Straight line is $x = $ constant, a line parallel to y-axis

The multiplication of an equation by a constant does not alter it.

If the equations $a_1 x + b_1 y + c_1 = 0$ and $A_1 x + B_1 y + C_1 = 0$ represent the same line, we must have

$$\frac{a_1}{A_1} = \frac{b_1}{B_1} = \frac{C_1}{C_1^1}$$

11. **Let us now determine the equation to the straight line which passes through the two given points (x', y') and (x'', y'').**

    The equation to any straight line is
    $$y = mx + C \quad (1)$$

    If this equation passes through the point (x', y') then
    $$y' = mx' + c$$

    Or $c = y' - mx'$ (2)

    (2) in (1) gives
    $$y = mx + y' - mx'$$

    Or $y - y' = m(x - x')$ (3)

    This represents equation of straight line passing through (x', y') and making angle $\tan^{-1} m$ with OX.

    If equation (3) passes through the point (x'', y'') then
    $$y'' - y' = m(x'' - x') \Rightarrow m = \frac{y'' - y'}{x'' - x'}$$

    Substituting this in (3), we get the required ex
    $$y - y' = \frac{y'' - y'}{x'' - x'}(x - x')$$

12. **Let us now determine the angles between straight lines given the general equation of these straight lines**

    Let the two straight lines be AL_1 and AL_2, meeting the axis of x in L_1 and L_2.

    Let their equations be
    $$y = m_1 x + c_1 \text{ and } y = m_2 x + c_2 \quad (1)$$

We therefore have

$$\tan \lfloor AL_1 = m_1 \text{ and } \tan \lfloor AL_2 X = m_2$$

Now $\lfloor L_1 AL_2 = \lfloor AL_1 X - \lfloor AL_2 X$

$$\therefore \tan \lfloor L_1 AL_2 = \tan(\lfloor AL_1 X - \lfloor AL_2 X)$$

$$= \frac{\tan \lfloor AL_1 X - \tan \lfloor AL_2 X}{1 + \tan \lfloor AL_1 X \tan \lfloor AL_2 X}$$

$$= \frac{m_1 - m_2}{1 + m_1 m_2}$$

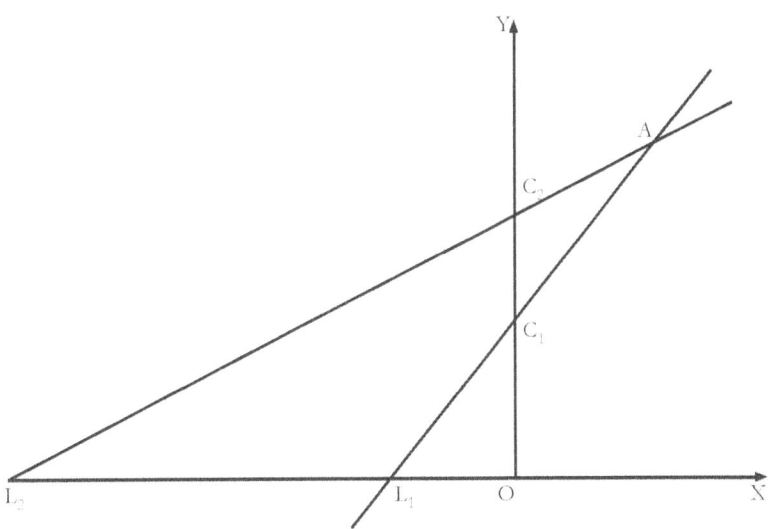

Fig. 33

Hence the required angle is

$$\lfloor L_1 AL_2 = \tan^{-1} \left(\frac{m_1 - m_2}{1 + m_1 m_2} \right) \quad (2)$$

13. Let us now determine the condition for the two straight lines to be parallel to one another

Two straight lines are parallel when the angle between them is zero and therefore the tangent of this angle is zero

$$\tan\theta = \frac{m_1 - m_2}{1 + m_1 m_2} = 0$$

$$\Rightarrow m_1 - m_2 = 0$$

Or $m_1 = m_2$

14. Let us now determine the condition for two straight lines to be perpendicular to one another.

Let the straight lines be

$$y = m_1 x + c_1$$

And $y = m_2 x + c_2$

If θ be the angle between them

Then $\tan\theta = \dfrac{m_1 - m_2}{1 + m_1 m_2}$ (1)

If the lines are perpendicular, then $\theta = 90°$

$$\therefore \tan\theta = \tan 90$$

$$= \infty$$

$\therefore$ (1) becomes

$$\infty = \frac{m_1 - m_2}{1 + m_1 m_2}$$

$$\frac{1}{0} = \frac{m_1 - m_2}{1 + m_1 m_2}$$

Or $1 + m_1 m_2 = 0$

Or $m_1 m_2 = -1$

The straight line $y = m_2 x + c_2$ is $\perp$ to $y = m_1 x + c_1$

If $m_2 = \dfrac{-1}{m_1}$

If $A_1 x + B_1 y + c_1 = 0$ & $A_2 x + B_2 y + c_2 = 0$ are perpendicular

Then $m_1 = \dfrac{-A_1}{B_1}$, $m_2 = \dfrac{-A_2}{B_2}$

Then $m_1 m_2 = -1$

$$\left(\frac{-A_1}{B_1}\right)\left(\frac{-A_2}{B_2}\right) = -1$$

Or $A_1 A_2 = -B_1 B_2$

Or $A_1 A_2 + B_1 B_2 = 0$

The two straight lines

$$A_1 x + B_1 y + C_1 = 0 \quad (1)$$

$$B_1 x - A_1 y + C_2 = 0 \quad (2)$$

Are at right angles, then

$$m_1 = \frac{-A_1}{B_1} \quad m_2 = \frac{B_1}{A_1}$$

$$\therefore m_1 m_2 = \frac{-A_1}{B_1} \times \frac{B_1}{A_1} = -1$$

$$m_1 m_2 = -1$$

Also (2) is derived from (1) by interchanging the coefficients of x and y, changing the sign of one them and changing the constant into any other constant.

15. **Let us now determine the equations to the straight lines which pass through a given point** (x', y') **and make a given angle** α **with the given straight line** $y = mx + C$

Let P be the given point and let the given straight line be LMN, making an angle θ with the axis of x such that $\tan \theta = m$.

In general there are two straight lines PMR and PNS making an angle α and α' with the given line.

Let these lines meet the axis of x in R and S and let them make angles ϕ with the positive direction of the axis of x.

The equations to the two required straight lines are therefore

$$y - y' = \tan\phi(x - x') \quad (1)$$

And

$$y - y' = \tan\phi'(x - x') \quad (2)$$

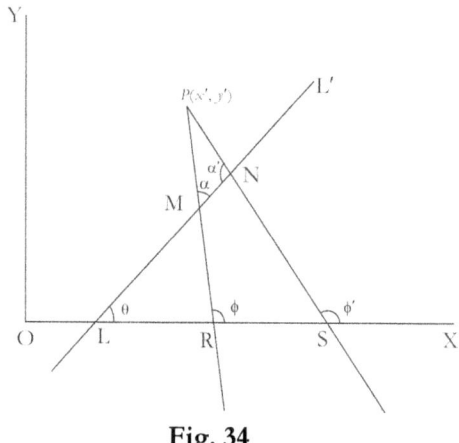

Fig. 34

Now $\phi = \underline{LMR} + \underline{RLM} = \alpha + \theta$

$$\& \, \phi' = \underline{LNS} + \underline{SLN} = (180° - \alpha) + \theta$$

Hence, $\tan\phi = \tan(\alpha + \theta) = \dfrac{\tan\alpha + \tan\theta}{1 - \tan\alpha\tan\theta}$

$$\tan\phi = \dfrac{\tan\alpha + m}{1 - m\tan\alpha}$$

And $\tan\phi' = \tan(180° + \theta - \alpha)$

$$= \tan(\theta - \alpha') = \dfrac{\tan\theta - \tan\alpha}{1 + \tan\theta\tan\alpha}$$

$$\therefore \tan \phi' = \frac{m - \tan \alpha}{1 + m \tan \alpha}$$

Substituting these values in (1) & (2), we get required equations

$$y - y' = \frac{\tan \alpha + m}{1 - m \tan \alpha}(x - x')$$

And

$$y - y' = \frac{m - \tan \alpha'}{1 + m \tan \alpha'}(x - x')$$

16. **Let us determine a way to show that the point (x', y') is on one side or the other of the straight line $Ax + By + C = 0$ according as the quantity $Ax' + By' + C$ is positive or negative.**

Let LM be the given straight line and P any point (x, y')

Through P draw PQ parallel to the axis of y, to meet the given straight line in Q, and let the co-ordinates of Q be (x', y'')

Since Q lies on the given line, we have

$$Ax' + By'' + C = 0$$

So that $y'' = -\dfrac{(C + Ax')}{B}$ (1)

It is clear from the figure that PQ is drawn parallel to the positive or negative direction of the axis of y according as P is on one side, or the other, of the straight line LM, i.e. according as y'' is > or < y'

i.e. $y'' - y'$ is positive or negative

Now by (1)

$$y'' - y' = -\frac{Ax' + C}{B} - y'$$

$$= -\frac{1}{B}[Ax' + By' + C]$$

The point (x', y') is therefore, on one side or the other of LM according as the quantity $Ax' + By' + C$ is negative or positive.

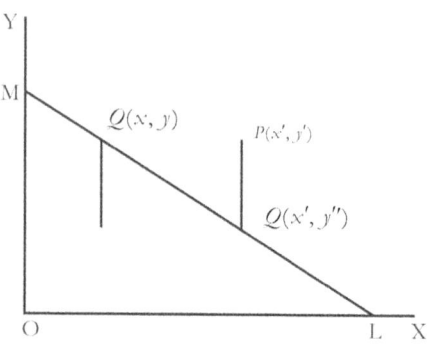

Fig. 35

17. **Let us now determine the length of a perpendicular let fall from a given point upon a given straight line**

   (i) Let the equation of straight line be
   $x\cos\alpha + y\sin\alpha - P = 0$ (1)

   Where P is the length of $\perp^{lar}$ on it

   We have ON = P & $\underline{XON} = \alpha$

   Let the given point P be (x', y)

   Through P draw PR parallel to the given line to meet ON produced in R and draw PQ the required $\perp^{lar}$.

   If OR be P' the equation to PR is

   $$x\cos\alpha + y\sin\alpha - P' = 0 \quad (2)$$

   This straight line passes through the point $P(x', y')$, we have

$$x'\cos\alpha + y'\sin\alpha - p' = 0$$

Or $p' = x'\cos\alpha + y'\sin\alpha$

But the required perpendicular

$$= PQ = NR = OR - ON$$

$$= P' - P$$

$$= x'\cos\alpha + y'\sin\alpha - P \ (3)$$

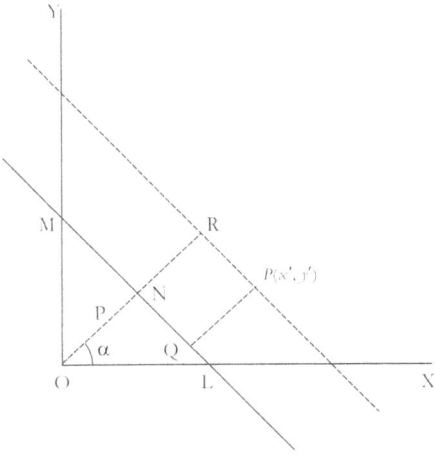

Fig. 36

The length of the required perpendicular is therefore, obtained by substituting x', and y' for x and y in the given equation.

(ii) Let the equation to the straight line be

$$Ax + By + C = 0 \ (4)$$

The equation being written so that C is a negative quantity

This equation is reduced to the standard form $x\cos\alpha + y\sin\alpha = P$ by dividing it by $\sqrt{A^2 + B^2}$. It the becomes

$$\frac{Ax}{\sqrt{A^2 + B^2}} + \frac{By}{\sqrt{A^2 + B^2}} + \frac{C}{\sqrt{A^2 + B^2}} = 0$$

Hence
$$\cos\alpha = \frac{A}{\sqrt{A^2 + B^2}}, \sin\alpha = \frac{B}{\sqrt{A^2 + B^2}} \text{ and}$$

$$-P = \frac{C}{\sqrt{A^2 + B^2}}.$$

The perpendicular from the point (x', y') therefore

$$= x'\cos\alpha + y'\sin\alpha - p$$

$$= \frac{Ax' + By' + C}{\sqrt{A^2 + B^2}}$$

Corolla: The perpendicular from origin $= \dfrac{C}{\sqrt{A^2 + B^2}}$

18. Let us determine the co-ordinates of the point of intersection of two given straight lines

Let the equations of the two straight lines be

$$a_1 x + b_1 y + c_1 = 0 \quad (1)$$

And

$$a_2 x + b_2 y + c_2 = 0 \quad (2)$$

And let the straight lines be AL_1 & AL_2.

Since (1) is the equation to AL_1, the co-ordinates of any point on it must satisfy (1). So the co-ordinates of any point on AL_2 satisfy (2).

Now the only point which is common to these two straight lines is their point of intersection A.

The coordinates of this point must therefore, satisfy both (1) and (2)

If therefore A be the point (x_1, y_1) we have

$$a_1 x_1 + b_1 y_1 + c_1 = 0 \quad (3)$$
$$a_2 x_1 + b_2 y_2 + c_2 = 0 \quad (4)$$

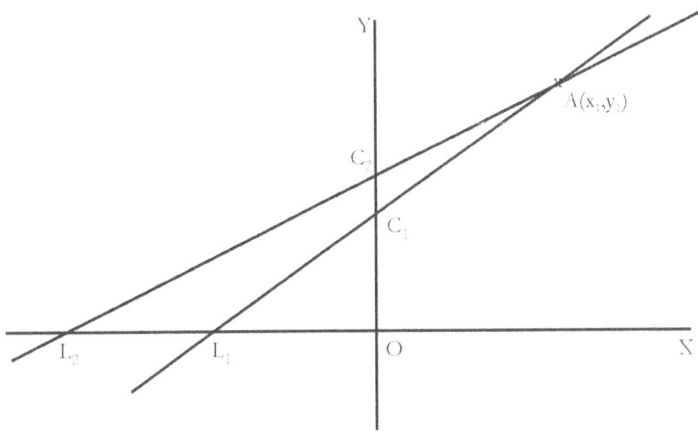

Fig. 37

Solving we have

$$\begin{array}{cccc} x & y & 1 \\ b_1 & c_1 & a_1 & b_1 \\ b_2 & c_2 & a_2 & b_2 \end{array}$$

$$\frac{x_1}{b_1 c_2 - b_2 c_1} = \frac{y_1}{c_1 a_2 - c_2 a_1} = \frac{1}{a_1 b_2 - a_2 b_1}$$

Or $x_1 = \dfrac{b_1 c_2 - b_2 c_1}{a_1 b_2 - a_2 b_1}, \; y_1 = \dfrac{c_1 a_2 - c_2 a_1}{a_1 b_2 - a_2 b_1}$

19. **Let us determine the condition that three straight lines may meet in point.**

Let the equations be

$$a_1 x + b_1 y + c_1 = 0 \quad (1)$$

$$a_2 x + b_2 y + c_2 = 0 \quad (2)$$

$$a_3 x + b_3 y + c_3 = 0 \quad (3)$$

If the three straight lines meet in a point let it be (x_1, y_1). So that the values x_1 and y_1 satisfy the equations (1), (2) & (3) and hence

$$a_1 x_1 + b_1 y_1 + c_1 = 0$$
$$a_2 x_1 + b_2 y_1 + c_2 = 0$$
$$a_3 x_1 + b_3 y_1 + c_3 = 0$$

The condition that these three equations should hold between the two quantities $x_1 \ y_1$ is

$$\begin{vmatrix} a_1 & b_1 & c_1 \\ a_2 & b_2 & c_2 \\ a_3 & b_3 & c_3 \end{vmatrix} = 0$$

20. Let us determine the equations of the bisectors of the angles between the straight lines

Let the two straight lines be represented by the following equations:

$$a_1 x + b_1 y + c_1 = 0 \quad (1)$$
$$a_2 x + b_2 y + c_2 = 0 \quad (2)$$

Let the straight lines be AL_1 and AL_2, and let the bisectors of the angles between then be AM_1 and AM_2

Let P be any point on either of these bisectors and draw PN_1 and $PN_2 \perp$ to the given lines

The $\Delta^{les} PAN_1$ and PAN_2 are equal in all respects so that the $\perp^{lass} PN_1$ and PN_2 are equal in magnitude.

Let the equations to the straight lines be written so that C_1 and C_2 are both negative, and to the quantities $\sqrt{a_1^2 + b_1^2}$ and $\sqrt{a_2^2 + b_2^2}$ let the positive sign be prefixed.

If P be the point (h, k), the numerical values of PN_1 & PN_2 are

$$\frac{a_1h + b_1k + c_1}{\sqrt{a_1^{\,2} + b_1^{\,2}}} \quad \& \quad \frac{a_2h + b_2k + c_2}{\sqrt{a_2^{\,2} + b_2^{\,2}}} \quad (1)$$

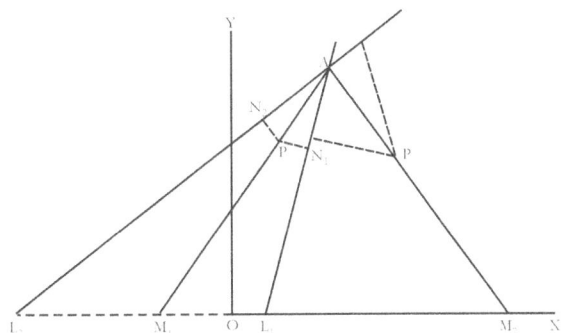

Fig. 38

If P lies on AM, on the bisector of the angle between the two straight lines in which the origin lies, the point P and the origin lie on the same side of each of the two lines

Hence the two quantities (1) have the same signs as C_1 and C_2 respectively

In this case, since C_1 and C_2 have the same sign, the quantities (1) have the same sign, and hence

$$\frac{a_1h + b_1k + c_1}{\sqrt{a_1^{\,2} + b_1^{\,2}}} = +\frac{a_2h + b_2k + c_2}{\sqrt{a_2^{\,2} + b_2^{\,2}}}$$

But this is the condition that the point (h, k) way lie on the straight line

$$\frac{a_1x + b_1y + c_1}{\sqrt{a_1^{\,2} + b_1^{\,2}}} = -\frac{a_2x + b_2y + c_2}{\sqrt{a_2^{\,2} + b_2^{\,2}}}$$

The equation to the original lines being therefore, arranged so that the constant terms are both positive(or −ve) then equation to the bisector is

$$\frac{a_1x + b_1y + c_1}{\sqrt{a_1^2 + b_1^2}} = \pm \frac{a_2x + b_2y + c_2}{\sqrt{a_2^2 + b_2^2}}$$

21. Let us determine the equation of straight line, which passes through a given point and makes a given angle θ with a given line,

Let A be the given point (h, k) and $L'AL$ a straight line through it inclined at an angle θ to the axis of x.

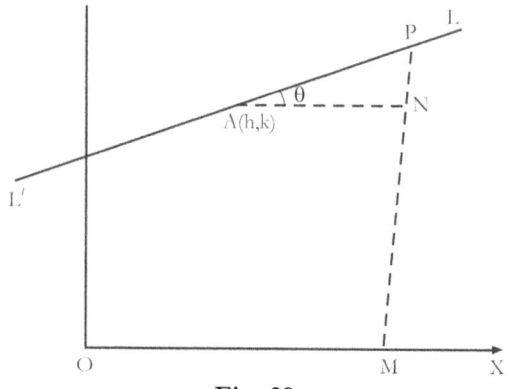

Fig. 39

Take any point P, whose co-ordinate are (x, y) lying on this line and let the distance AP be r

Draw $PM \perp$ to axis of x and $AN \perp$ to PM

Then

$$x - h = AN = AP\cos\theta = r\cos\theta$$
$$y - k = NP = AP\sin\theta = r\sin\theta$$

Hence $\dfrac{x - h}{\cos\theta} = \dfrac{y - k}{\sin\theta} = r$

Cor1: From above equation

$$x = h + r\cos\theta, \quad y = k + r\sin\theta$$

The co-ordinates of any point on the given line are therefore $h + r\cos\theta$, and $k + r\sin\theta$.

Solved Examples

Example 1:   Find the equation to the straight line passing through the point $(3, -4)$ and cutting off intercepts, equal but of opposite signs, from the two axes.

Let the equation of straight line be $\dfrac{x}{a} + \dfrac{y}{b} = 1$

Given $a = -b$ and it passes through the point $(3, -4)$

$$\therefore \frac{3}{-b} + \frac{(-4)}{b} = 1$$

$$\frac{3+4}{-b} = 1 \Rightarrow -b = 7 \text{ or } b = 7$$

$$a = -7$$

$\therefore$ required y is $\dfrac{x}{-7} + \dfrac{y}{7} = 1$

Or $x - y = 7$

Example 2:   Find the equation to the straight line which passes through the point $(-5, 4)$ and is such that the portion of it between the axes is divided by the point in the ratio of 1:2.

Let the required straight line be $\dfrac{x}{a} + \dfrac{y}{b} = 1$

This meets the axes in the points whose co-ordinates are $(a, 0) \,\&\, (0, b)$

The co-ordinates of the point dividing the line joining these points in the ratio 1:2 are

Internal division formula is

$$P(x, y) = \left(\frac{m_1 x_2 + m_2 x_1}{m_1 + m_2}, \frac{m_1 y_2 + m_2 y_1}{m_1 + m_2} \right) \quad \begin{array}{ll} A(a, 0) & B(0, b) \\ x_1 \, y_1 & x_2 \, y_2 \end{array}$$

$$= \left(\frac{1(0) + 2(a)}{1+2}, \frac{1(b)2(0)}{1+2} \right) \quad \begin{array}{l} m_1 : m_2 = 1 : 2 \\ \Rightarrow m_1 = 1, m_2 = 2 \end{array}$$

$$= \left(\frac{2a}{3}, \frac{b}{3} \right)$$

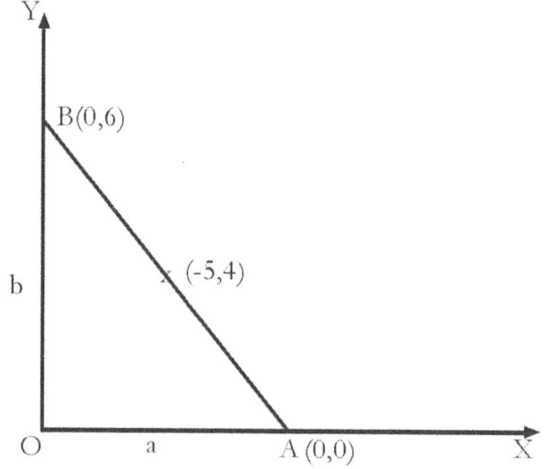

Fig 40

If this be the point $(-5, 4)$ we have

$$(-5, 4) = \left(\frac{2a}{3}, \frac{b}{3} \right)$$

$$-5 = \frac{2a}{3} \quad 4 = \frac{b}{3}$$

Or $a = \dfrac{-15}{2}$ or $b = 12$

The required straight line is therefore

$$\frac{x}{\dfrac{-15}{2}} + \frac{y}{12} = 1$$

Or $\dfrac{-8x + 5y}{60} = 1$

Or $8x - 5y + 60 = 0$

Example 3: Trace the straight lines: $3x - 4y + 7 = 0$, $7x + 8y + 9 = 0$, $3y = x$, $x = 2$, $y = -2; x = 0$

1. $3x - 4y + 7 = 0$

Put $x = 0 \ 4y = 7 \Rightarrow y = \dfrac{7}{4} \ B\left(0, \dfrac{7}{4}\right)$

$y = 0 \Rightarrow 3x = -7$ or $x = \dfrac{-7}{3} \ A\left(\dfrac{-7}{3}, 0\right)$

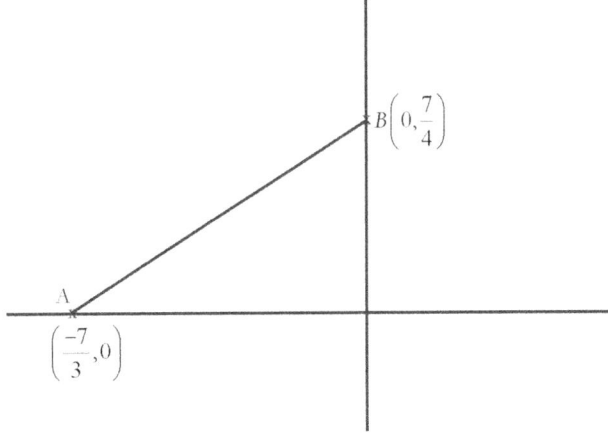

Fig. 41

2. $7x + 8y + 9 = 0$

Put $x = 0$, then $y = \dfrac{-9}{8}$ hence $B\left(0, \dfrac{-9}{8}\right)$

Put $y = 0$, then $x = \dfrac{-9}{7}$ hence $A\left(\dfrac{-9}{7}, 0\right)$

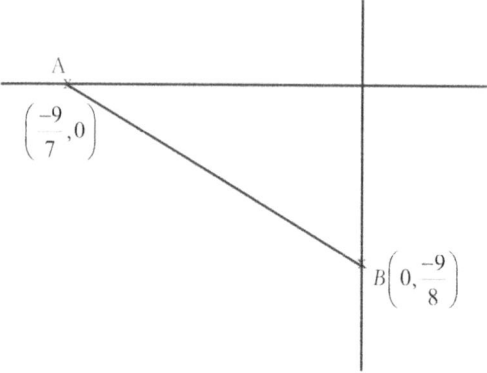

Fig. 42

3. $3y = x$

Put $x = 0$ $y = 0$ $B(0,0)$

Put $x = 3$ we get $y = 1$. $C(3,1)$

Connecting these two points we get the straight line defined by $3y = x$

4. $x = 2$ $A(2,0)$

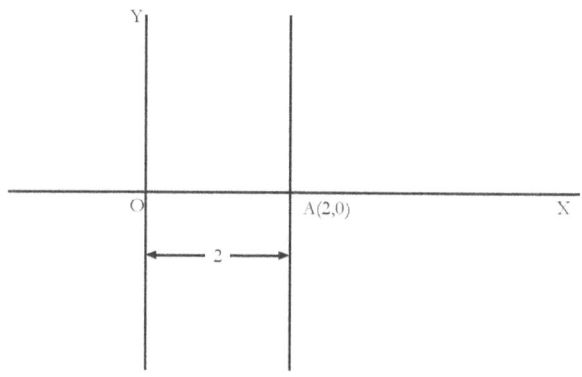

Fig. 43

5. $y = -2$ $x = 0$ $B(0,-2)$

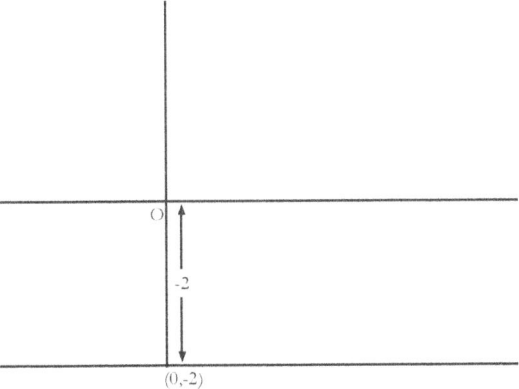

Fig. 44

Example 4: Find the equation to the straight line which passes through the points $(-1, 3)$ and $(4, -2)$

We know that equation of straight line passing through two points is

$$y - y' = \frac{y'' - y'}{x'' - x'}(x - x') \quad \begin{matrix} (-1, 3) \\ x' \; y' \end{matrix} \; \& \; \begin{matrix} (4_1 - 2) \\ x'' \; y'' \end{matrix}$$

$$(y - 3) = \frac{-2 - 3}{4 - (-1)}(x - (-1))$$

$$(y - 3) = \frac{-5}{5}(x + 1)$$

Or $y - 3 = -(x + 1)$

Or $x + y - 2 = 0$

Example 5: Find the equation to the straight line, which passes through the points $(4, -5)$ and which is parallel to the straight line $3x + 4y + 5 = 0$

Any straight line which is parallel to (1) is of the form
$$3x + 4y + C = 0 \quad (2)$$

Given this equation passes through $(4, -5)$

59

$$\therefore 3(4)+4(-5)+C=0$$
$$12-20+C=0$$
$$-8+C=0$$
$$C=8$$
$\therefore$ equation (2) becomes $3x+4\,y+8=0$

Example 6:   Find the equation to the straight line which passes through the point $(4,-5)$ and is perpendicular to the straight line $3x+4\,y+5=0$

Any straight line perpendicular to $3x+4\,y+5=0$ is of the form
$$4x-3\,y+C=0 \ (2)$$
(2) passes through the point $(4,-5)$
$$4(4)-3(-5)+C=0$$
$$16+15+C=0$$
Or $C=-31$
$\therefore$ Required equation is $4x-3\,y=31$

Example 7:   Find the equation to the straight line which passes through the point (x',y') and is perpendicular to the given straight line $y\,y'=2a(x+x')$

The given straight line is $y\,y'-2ax-2ax'=0$

Any straight line $\perp^{lae}$ to it is $2ay+xy'+C=0$ (1)

This will pass through the point (x',y') and therefore, will be the straight line required if the co-ordinates x' & $4\,y'$ satisfy it.

If $2ay'+x'y'+C=0$

If $C=-2ay'-x'y'$

Substituting in (1) for C the required equation is therefore

$$2a(y-y')+y'(x-x')=0$$

Example 8:   Show that the three straight lines $2x-3y+5=0$,
$3x+4y-7-0$ and $9x-5y+8=0$ meet in a point.

We know that if three lines meet in a point

$$\begin{vmatrix} a_1 & b_1 & c_1 \\ a_2 & b_2 & c_2 \\ a_3 & b_3 & c_3 \end{vmatrix} = 0$$

i.e. $\begin{vmatrix} 2 & -3 & 5 \\ 3 & 4 & -7 \\ 9 & -5 & 8 \end{vmatrix}$

$$= 2(32-35)+3(24+63)+5(-15-36)$$
$$= 2(-3)+3(87)+5(-51)$$
$$= -6+261-255$$
$$= -261+261$$
$$= 0$$

Example 9:   Prove that the three perpendiculars drawn from the vertices of a triangle upon the opposite sides all meet in a point.

Let the triangle be ABC and let its angular points be

$$(x_1,y_1),(x_2,y_2) \ \& \ (x_3,y_3)$$

The equation to BC is $(y-y_2)=\dfrac{y_3-y_2}{x_3-x_2}(x-x_2)$

The equation to the perpendicular from A on this straight line is

$$y-y_1=-\left(\frac{x_3-x_2}{y_3-y_2}\right)(x-x_1)$$

i.e. $(y-y_1)(y_3-y_2)=-(x_3-x_2)(x-x_1)$

Or $y(y_3-y_2)+x(x_3-x_2)=y_1(y_3-y_2)+x_1(x_3-x_2)$ (1)

So, the perpendiculars from B and C on CA & AB are

$$y(y_1 - y_3) + x(x_1 - x_3) = y_2(y_1 - y_3) + x_2(x_1 - x_3)$$
(2)

And

$$y(y_2 - y_1) + x(x_2 - x_1) = y_3(y_2 - y_1) + x_3(x_2 - x_1)$$
(3)

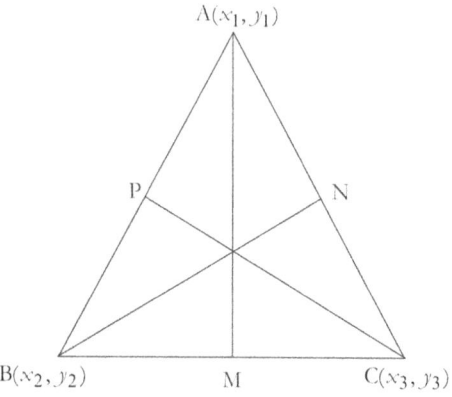

Fig. 45

On adding these three equations their sum identically vanishes. The straight lines represented by them therefore meet in a point. This point is called the orthocenter of the triangle.

Note: To find the equation to any straight line which passes through the intersection of the two straight lines

$$a_1 x + b_1 y + c_1 = 0 \quad (1)$$

$$a_2 x + b_2 y + c_2 = 0 \quad (2)$$

If (x_1, y_1) be the common point of the equations (1) & (2) we find the values of x_1, y_1, and then the equation to any straight line through it is $y - y_1 = m(x - x_1)$, where m is a constant.

Example 10: Find the equation to the straight line which passes through the intersection of the straight lines $2x - 3y + 4 = 0$,

$3x + 4y - 5 = 0$ (1)   and   is   perpendicular   to   the   straight   line
$6x - 7y + 8 = 0$

Solving (1) for x_1, y_1

$$
\begin{array}{cccc}
x_1 & y_1 & 1 \\
-3 & 4 & 2 & -3 \\
4 & -5 & 3 & 4
\end{array}
$$

$$\frac{x_1}{15 - 16} = \frac{y_1}{12 + 10} = \frac{1}{8 + 9}$$

$$x = \frac{-1}{17}, \; y = \frac{22}{17}$$

The equation of any straight line through this common point
is therefore

$$y - \frac{22}{17} = m\left(x + \frac{1}{17}\right)$$

$$\therefore m \times \frac{6}{7} = -1$$

Or $m = \dfrac{-7}{6}$

The required equation is therefore

$$y - \frac{22}{17} = \frac{-7}{6}\left(x + \frac{1}{17}\right)$$

$$\frac{17y - 22}{\cancel{17}} = \frac{-7}{6}\left(\frac{17x + 1}{\cancel{17}}\right)$$

$$6(17y - 22) = -7(17x + 1)$$

Or $119x + 102y = 125$

Alternatively, we can solve this in this method also.

This straight line is $\perp^{lae}$ to $6x - 7y + 8 = 0 \Rightarrow$ equation of required line is $7x + 6y + c = 0$

In this equation, we ubstitute the value $\left(\dfrac{-1}{17}, \dfrac{22}{17}\right)$

$$\dfrac{-7}{17} + \dfrac{132}{17} + c = 0$$

$$c = \dfrac{-125}{17}$$

$\therefore$ Equation in $7x + 6y \dfrac{-125}{17} = 0$

$$119x + 102y = 125$$

Example 11: Given the vertical angle of a triangle in magnitude and position, and also the sum of the reciprocals of the sides which contain it. Show that the base always passes through a fixed point.

Take the fixed angular point as origin and the directions of the sides containing it as axes, let the lengths of these sides in any such triangle be a and b, which are not therefore given

We have $\dfrac{1}{a} + \dfrac{1}{b} = \text{constant} = \dfrac{1}{k}$ (say)

The equation to the base is

$$\dfrac{x}{a} + \dfrac{y}{b} = 1$$

i.e. by (1) $\dfrac{x}{a} + y\left(\dfrac{1}{k} - \dfrac{1}{a}\right) = 1$

$$\because \dfrac{1}{b} = \dfrac{1}{k} - \dfrac{1}{a}$$

$$\dfrac{1}{a}(x - y) + \dfrac{y}{k} - 1 = 0$$

Whatever be the value of a this straight line always passes through the point given by

$$x - y = 0 \text{ and } \frac{y}{k} - 1 = 0$$

$$x = y \quad y - k = 0$$

$$y = k$$

$$\therefore x = y = k$$

i.e. through the fixed point (k, k).

Example 12: Prove that the co-ordinates of the centre of the circle inscribed in the triangle, whose vertices are the point $(x_1, y_1), (x_2, y_2)$ and (x_3, y_3) are $\dfrac{ax_1 + bx_2 + cx_3}{a + b + c}$ and $\dfrac{ay_1 + by_2 + cy_3}{a + b + c}$ Where a, b and c are the lengths of the sides of the triangle. Find also the co-ordinate of the centers of the ascribed circle.

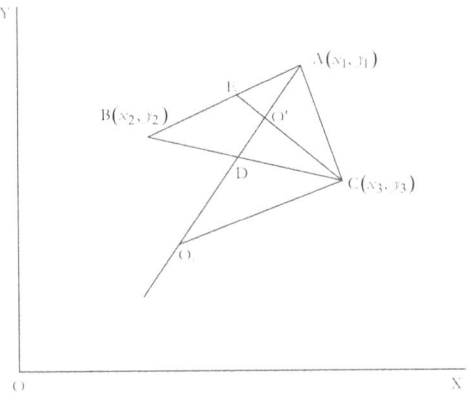

Fig. 46

Let ABC be the triangle and let AD and CE be the bisectors of the angles A and C and let them meet in O'.

The O' is the required point.

Since, AD bisects the angle BAC we have, by geometry

$$\frac{BD}{BA} = \frac{DC}{AC} = \frac{BD+DC}{BA+AC} = \frac{a}{b+c}$$

$$\because AC = b$$

$$BC = a = BD + DC$$

$$AB = C$$

So that $DC = \dfrac{ba}{b+c}$

Also, since CO' bisects the angle ACD, we have

$$\frac{AO'}{O'D} = \frac{AC}{CD} = \frac{\dfrac{b}{ba}}{b+c} = \frac{b+c}{a}$$

The point D therefore, divides BC in the ratio $c : b$

$$: BD : DC = c : b$$

Also O' divides AD in the ratio $b + c : a$

We have any point $P(x, y)$ divides $A(x_1, y_1)$ & $B(x_2, y_2)$ internally in the ratio $m : n$ then

$$P(x, y) = \left(\frac{mx_2 + nx_1}{m+n}, \frac{my_2 + ny_1}{m+n} \right)$$

The co-ordinates of D are

$$m : n = c : b$$

$$(x_1, y_1) = B(x_2 \, y_2)$$

$$(x_2, y_2) = C(x_3 \, y_3)$$

$$= \left(\frac{cx_3 + bx_2}{c+b}, \frac{cy_3 + by_2}{c+b} \right)$$

Again the co-ordinates of O' are

$$b + c : a = m : n$$

$$O' = \left(\frac{(b+c)\left(\dfrac{cx_3 + bx_2}{c+b} \right) + ax_1}{(b+c)+a}, \frac{(b+c)\left(\dfrac{cy_3 + by_2}{c+b} \right) + ay_1}{(b+c)+a} \right)$$

$$= \left(\frac{cx_3 + bx_2 + ax_1}{a+b+c}, \frac{cy_3 + by_2 + ay_1}{a+b+c} \right)$$

$$O' = \left(\frac{ax_1 + bx_2 + cx_3}{a+b+c}, \frac{ay_1 + by_2 + cy_3}{a+b+c} \right)$$

Again, if O_1 is the center of the escribed circle opposite to the angle A, the line CO_1 bisects the exterior angle of ACB.

Hence, by geometry, we have

$$\frac{AO_1}{O_1 D} = \frac{AC}{CD} = \frac{b+c}{a}$$

Therefore, O_1 is the point which divides AD externally in the ration $b+c : a$.

Its co-ordinates are therefore

$$= \left(\frac{(b+c)\dfrac{cx_3 + bx_2}{c+b} - ax_1}{b+c-a}, \frac{(b+c)\left(\dfrac{cy_3 + by_2}{c+b} \right) - ay_1}{b+c-a} \right)$$

$$= \left(\frac{-ax_1 + bx_2 + cx_3}{-a+b+c}, \frac{-ay_1 + by_2 + cy_3}{-a+b+c} \right)$$

Similarly, it may be shown that the coordinates of the centers of the escribed circles opposite toe B and C are respectively.

$$\left(\frac{ax_1 - bx_2 + cx_3}{a-b+c}, \frac{ay_1 - by_2 + cy_3}{a-b+c} \right)$$

And $\left(\dfrac{ax_1 + bx_2 - cx_3}{a+b-c}, \dfrac{ay_1 + by_2 - cy_3}{a+b-c} \right)$

Example 13: Prove that the middle points of the diagonals of a complete quadrilateral lie on the same straight line.

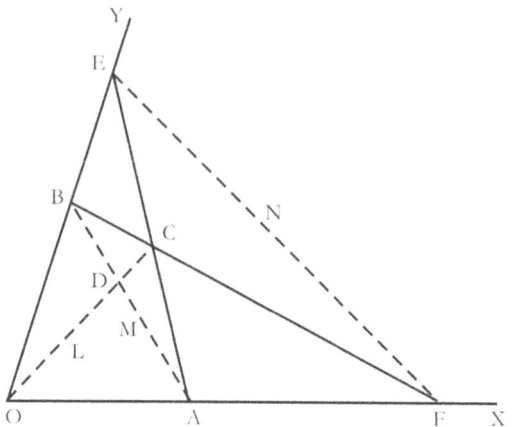

Fig. 47

Let OACB be any quadrilateral

Let AC and OB be produced to meet in E, and BC and OA to meet in F. Join AB, OC & EF.

The resulting figure is called "complete quadrilateral", and the points E, F & D (the intersection of AB, and OC) are called it vertices.

Take the lines OAF and OBE as the axes of x & y.

Let $OA = 2a$ and $OB = 2b$, so that A is the point $(2a, 0)$ and B is the point $(0, 2b)$, also let C be the point $(2h, 2k)$.

Then L is the middle point of OC, is the point (h, k) and M, the middle point of AB is (a, b).

The equation to LM is therefore,

$$y - b = \frac{k-b}{h-a}(x-a)$$

I.e. $(h-a)(y-b) = (k-b)(x-a)$

Or $(h-a)y - (k-b)x = ay - ab - ak + ab$

$$(b-a)y - (k-b)x = ay - ak \qquad\qquad (1)$$

Again, the equation to BC is $y - 2b = \dfrac{k-b}{b}x$

Putting $y = 0$, we have $x = \dfrac{-2bb}{k-b}$, so that F is the point

$\left(\dfrac{-2bb}{k-b}, 0\right)$

Similarly, E is the point $\left(0, \dfrac{-2ak}{b-a}\right)$

Hence, N the middle point of EF is $\left(\dfrac{-bb}{k-b}, \dfrac{-ak}{b-a}\right)$

These co-ordinates clearly satisfy (1) i.e. N lies on the straight line LM.

Example 14: The base of a triangle is fixed, find the locus of the vertex when one base angle is double of the other.

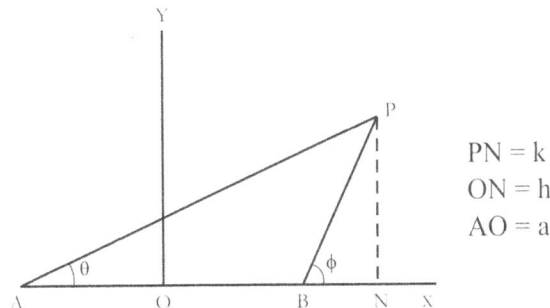

PN = k
ON = h
AO = a

Fig. 48

Let AB be the fixed base of the triangle, take its middle point O as origin, the direction of OB as the axis of x and a $\perp^{lar}$ line as the axes of y.

Let $AO = OB = a$

If P be one position of the vertex, the condition of the problem then gives $p(h, k)$

$$\lfloor PBA = 2 \lfloor PAB$$

i.e. $\pi - \phi = 2\theta$

i.e. $\tan(\pi - \phi) = \tan 2\theta$

Or $-\tan \phi = \tan 2\theta$ (1)

$$\left(\because \tan 2\theta = \frac{2\tan\theta}{1 - \tan^2\theta} \right)$$

We then have

$$\frac{K}{h + a} = \tan\theta \text{ and } \frac{K}{h - a} = \tan\phi$$

Substituting these values in (1), we have

$$\frac{-K}{h - a} = \frac{\dfrac{2k}{h + a}}{1 - \left(\dfrac{K}{h + a}\right)^2}$$

$$\text{Or} \frac{-K}{h - a} = \frac{\dfrac{2k}{h + a}}{\dfrac{h^2 + a^2 + 2ah - k^2}{(h + a)^2}}$$

$$\frac{-k}{h - a} = \frac{2k(h + a)}{h^2 + a^2 + 2ah - k^2}$$

$$-h^2 - a^2 - 2ah + k^2 = 2(h^2 - a^2)$$

Or $K^2 - 3h^2 - 2ah + a^2 = 0$

But this is the condition that the point (h, k) should lie on the curve

$$y^2 - 3x^2 - 2ax + a^2 = 0$$

This is therefore the equation to the required locus.

Example 15: A variable straight line is drawn through a given point O to cut two fixed straight lines in R and S; on it is taken a point P such that $\dfrac{2}{OP} = \dfrac{1}{OR} + \dfrac{1}{OS}$. Show that the locus of P is a third straight line.

Take any two fixed straight lines, at right angles and passing through O, as the axes and let the equation to the two given fixed straight lines be

$$Ax + By + C = 0 \text{ and } A'x + B'y + C' = 0$$

Transforming to polar co-ordinates these equations are

$$\frac{1}{r} = -\frac{A\cos\theta + B\sin\theta}{C} \text{ and } \frac{1}{r} = -\frac{A'\cos\theta + B'\sin\theta}{C'}$$

If the angle XOR be θ the values of $\dfrac{1}{OR}$ and $\dfrac{1}{OS}$ are therefore

$$\frac{1}{OR} + \frac{1}{OS} = -\frac{A\cos\theta + B\sin\theta}{C} - \frac{A'\cos\theta + B'\sin\theta}{C}$$

$$= \frac{-1}{r} - \frac{1}{r}$$

$$\frac{-2}{r} = -\left(\frac{A}{C} + \frac{A'}{C'}\right)\cos\theta - \left(\frac{B}{C} + \frac{B'}{C'}\right)\sin\theta$$

The equation to the locus of P is therefore, on again transforming to Cartesian co-ordinates

$$2 = -x\left(\frac{A}{C} + \frac{A'}{C'}\right) - y\left(\frac{B}{C} + \frac{B'}{C'}\right)$$

And this is a fixed straight line.

Example 16: Find the equations to the bisectors of the angles between the straight lines $3x - 4y + 7 = 0$ and $12x - 5y - 8 = 0$

Given $3x - 4y + 7 = 0$ & $-12x + 5y + 8 = 0$

The equation to the bisector of the angle in which the origin lies is therefore

$$\frac{3x - 4y + 7}{\sqrt{3^2 + 4^2}} = \frac{-12x + 5y + 8}{\sqrt{12^2 + 5^2}}$$

i.e. $\dfrac{3x - 4y + 7}{\sqrt{25}} = \dfrac{-12x + 5y + 8}{\sqrt{169}}$

Or $13(3x - 4y + 7) = 5(-12x + 5y + 8)$

Or $99x - 77y + 51 = 0$

The equation to the other bisector is

$$\frac{3x - 4y + 7}{\sqrt{3^2 + 4^2}} = -\frac{-12x + 5y + 8}{\sqrt{12^2 + 5^2}}$$

i.e. $13(3x - 4y + 7) + 5(-12x + 5y + 8) = 0$

Or $21x + 27y - 131 = 0$

5. The Straight Line: Polar Coordinates

We have discussed the various scenarios and problem formulations while dealing with a straight line in Cartesian rectangular coordinate system. We will now turn your attention to performing a few manipulations using polar coordinates.

1. **Let us commence our discussion by determining the general equation to straight line in polar co-ordinates:**

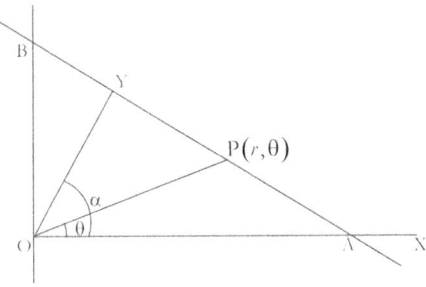

Fig. 49

Let P be the length of the perpendicular OY from the origin upon the straight line, let OY makes an angle α with the initial line.

$$\therefore \lfloor AOY = \alpha$$

Let P be any point on the line and let its co-ordinates be r and θ.

From $\triangle OYP$, we have

$$\cos YOP = \cos(\alpha - \theta) = \frac{OY}{OP} = \frac{p}{r}$$

$$\therefore p = r\cos(\alpha - \theta)$$

$$\therefore p = r\cos(\theta - \alpha)$$

This is the required equation.

2. **Let us now derive the polar equation of the straight line joining the points whose co-ordinates are** (r_1, θ_1) **and** (r_2, θ_2).

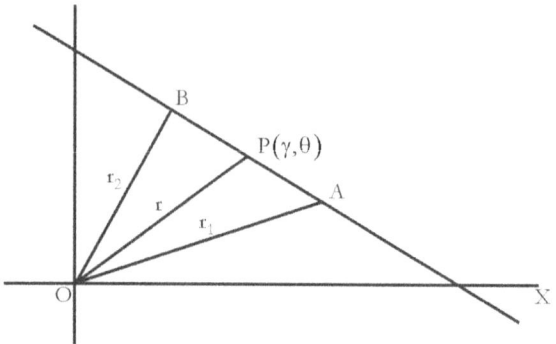

Fig. 50

Let A and B be the two given points and P any point on the line joining them. Whose co-ordinates are r and θ.

Then since

$$\triangle AOB = \triangle AOP + \triangle POB, \text{ we have}$$

$$\frac{1}{2} r_1 r_2 \sin AOB = \frac{1}{2} r_1 r \sin AOP + \frac{1}{2} r r_2 \sin POB$$

$$\because AOB = \theta_2 - \theta_1$$

$$AOP = \theta - \theta_1$$

$$POB = \theta_2 - \theta$$

Or $r_1 r_2 \sin(\theta_2 - \theta_1) = r_1 r \sin(\theta - \theta_1) + r r_2 \sin(\theta_2 - \theta)$

i.e. $\dfrac{\sin(\theta_2 - \theta_1)}{r} = \dfrac{\sin(\theta - \theta_1)}{r_2} + \dfrac{\sin(\theta_2 - \theta)}{r_1}$ $(\div \text{ by } r_1 r_2 r)$

Note: Oblique co-ordinates are not covered at middle school level. However, it is important for us to have an idea on how to deal with oblique coordinates.

3. **To find the equation to a straight line referred to axes inclined at angle w.**

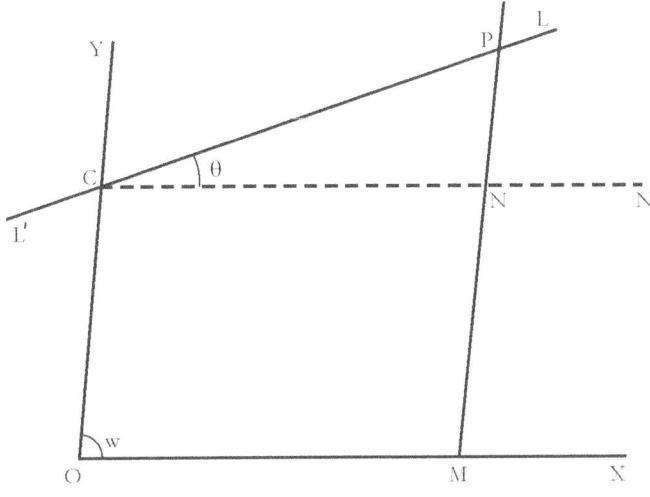

Fig. 51

Let LPL' be a straight line which cuts the axis of Y at a distance C from the origin and is inclined at an angle θ to the axis of x.

Let P be any point on the straight line. Draw PNM parallel to the axis of y to meet OX in M, and let it meet the straight line through C parallel to the axis of x in the point N.

Let P be the point (x, y), so that

$$CN = OM = x \text{ , and } NP = MP - OC = y - c$$

Since $\underline{|CPN} = \underline{|PNN'} - \underline{|PCN'} = w - \theta$, we have

$$\frac{y-c}{x} = \frac{NP}{CN} = \frac{\sin NCP}{\sin CPN} = \frac{\sin \theta}{\sin(w - \theta)}$$

Hence $y = x \dfrac{\sin \theta}{\sin(w - \theta)} + c$ (1)

This equation is of the form

$$y = mx + c$$

Where $m = \dfrac{\sin\theta}{\sin(w-\theta)} = \dfrac{\sin\theta}{\sin w\cos\theta - \cos w\sin\theta}$

$$= \dfrac{\sin\theta}{\cos\theta\left(\sin w - \cos w\dfrac{\sin\theta}{\cos\theta}\right)}$$

$$m = \dfrac{\tan\theta}{\sin w - \cos w\tan\theta}$$

And therefore, $\tan\theta = \dfrac{m\sin w}{1 + m\cos w}$

$(\because m\sin w - m\cos w\tan\theta = \tan\theta$

$m\sin w = \tan\theta(1 + m\cos w))$

$\therefore \tan\theta = \dfrac{m\sin w}{1 + m\cos w}$

In oblique co-ordinates the equation $y = mx + c$, therefore represent a straight line which is inclined at an angle $\tan^{-1}\left(\dfrac{m\sin w}{1 + m\cos w}\right)$ to the axis of x.

4. **Corollary: Let us determine the equation of a line perpendicular to the axes inclined at an angle w.**

From (1), we have $y = x\dfrac{\sin\theta}{\sin(w-\theta)} + c$

If $\theta = 90$ then $y = \dfrac{x\sin 90}{\sin(w - 90)} + c$

$$y = \dfrac{x}{-\cos w} + c$$

If this passes through origin then $0 = 0 + c \Rightarrow c = 0$

$$\therefore y = \dfrac{-x}{\cos w}$$

If $\theta = 90 + w$, then

$$y = \frac{x\sin(90+w)}{\sin(w-90-w)} + c$$

$$y = \frac{x\cos w}{-1} + c$$

$\because c = 0$ if pass through origin

$\therefore y = -x\cos w$

Equation to the straight line is $\perp$ to the axes of x & y is

$$y = \frac{-x}{\cos w} \text{ and } y = -x\cos w$$

5. **The axes being oblique, determine the equation to the straight line, such that the perpendicular on it from the origin is of length P and makes angles α and β with the axes of x & y.**

Let LM be the given straight line and OK the perpendicular on it from the origin.

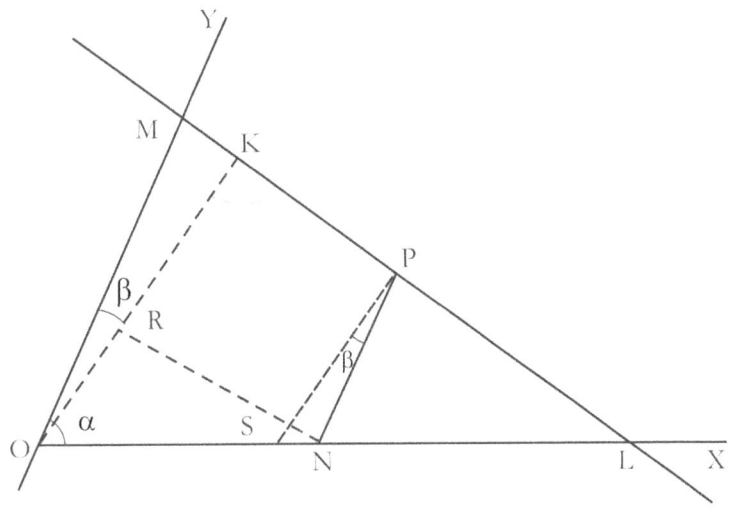

Fig. 52

Let P be any point on the straight line, draw the ordinate PN and draw $NR \perp$ to OK and $PS \perp$ to NR.

Let P be the point (x, y), so that

$$ON = x \text{ and } NP = y$$

The lines NP and OY are parallel.

Also OK & SP are parallel, each being $\perp$ to NR

Thus $\lfloor SPN = \lfloor KOM = \beta$

We therefore have

$$P = OK = OR + RK$$
$$= OR + SP$$
$$= ON\cos\alpha + NP\cos\beta$$
$$= x\cos\alpha + y\cos\beta$$

Hence $x\cos\alpha + y\cos\beta - p = 0$ being the relation which holds between the co-ordinates of any point on the straight line, is the required equation.

6. **Let us now determine the angle between two straight lines, when the axes are oblique.**

$y = mx + c$ and $y = m'x + c'$ the axes being oblique.

If these straight lines be respectively inclined at angle θ and θ' to the axis of x, then we have

$$\tan\theta = \frac{m\sin w}{1 + m\cos w} \text{ and } \tan\theta' = \frac{m'\sin w}{1 + m'\cos w}$$

The angle required is $\theta \sim \theta'$

$$\text{Consider} \tan(\theta - \theta') = \frac{\tan\theta - \tan\theta'}{1 + \tan\theta\tan\theta'}$$

$$= \frac{\dfrac{m\sin w}{1+m\cos w} - \dfrac{m'\sin w}{1+m'\cos w}}{1 + \dfrac{m\sin w}{1+m\cos w} \cdot \dfrac{m'\sin w}{1+m'\cos w}}$$

$$= \frac{m\sin w + mm'\sin w\cos w - m'\sin w - mm'\sin w\cos w}{1 + m\cos w + m'\cos w + mm'\cos^2 w + mm'\sin^2 w}$$

$$= \frac{\sin w(m - m')}{1 + \cos w(m + m') + mm'(1)}$$

$$\because \cos^2 w + \sin^2 w = 1$$

$$\tan(\theta - \theta') = \frac{(m - m')\sin w}{1 + (m + m')\cos w + mm'}$$

The required angle is therefore

$$\theta - \theta' = \tan^{-1}\left\{ \frac{(m - m')\sin w}{1 + (m + m')\cos w + mm'} \right\}$$

Corollary 1:– **The two given lines are parallel if** $m = m'$ **.ie.**
$\theta - \theta' = 0$

$$\therefore \tan 0 = 0 = \frac{(m - m')\sin w}{1 + (m + m')\cos w + mm'}$$

$$\Rightarrow (m - m')\sin w = 0$$

$$\Rightarrow m - m' = 0$$

$$\Rightarrow m = m'$$

As $m - m' \neq = 0$

$$\sin w = 0$$

Corollary 2:– If the two given lines are perpendicular

If $1 + (m + m')\cos w + mm' = 0$

i.e. $\theta - \theta' = 90°$

$$\therefore \tan 90° = \infty = \frac{1}{0} = \frac{(m-m')\sin w}{1+(m+m')\cos w + mm'}$$

$$\Rightarrow 1+(m+m')\cos w + mm' = 0$$

If the straight line have their equations in the form

$$Ax + By + C = 0 \text{ and } A'x + B'y + C' = 0$$

Then $m = \dfrac{-A}{B}$ and $m' = \dfrac{-A'}{B'}$

Substituting these values in

$$\theta - \theta' = \tan^{-1}\left\{\frac{(m-m')\sin w}{1+(m+m')\cos w + mm'}\right\}$$

We get

$$= \tan^{-1}\left\{\frac{\left(\dfrac{-A}{B}+\dfrac{A'}{B'}\right)\sin w}{1+\left(\dfrac{-A}{B}-\dfrac{A'}{B'}\right)\cos w + \left(\dfrac{-A}{B}\cdot\dfrac{-A'}{B'}\right)}\right\}$$

$$= \tan^{-1}\left\{\frac{\left(\dfrac{A'B-AB'}{BB'}\right)\sin w}{\dfrac{BB'-(AB'+BA')\cos w + AA'}{BB'}}\right\}$$

$$= \tan^{-1}\left\{\frac{(A'B-AB')\sin w}{AA'+BB'-(AB'+BA')\cos w}\right\}$$

The given lines are parallel if

$$A'B - AB' = 0$$

They are $\perp^r$ if $AA' + BB' = (AB' + BA')\cos w$

7. **Let us now determine the length of the perpendicular from the point** (x', y') **upon the straight line** $Ax + By + C = 0$, **the axes being inclined at an angle w, and the equation being written so that C is a negative quantity.**

Let the given straight line meet the axis in L and M, so that

$$OL = \frac{-C}{A} \text{ and } OM = \frac{-C}{B}$$

Let P be the point (x', y'). Draw the perpendiculars PQ, PR and PS on the given line and the two axes.

Taking O and P on opposite sides of the given line, we then have

$$\Delta LPM + \Delta MOL = \Delta OLP + \Delta OPM$$

i.e. $PQ \cdot LM + OL \cdot OM \sin w = OL \cdot PR + OM \cdot PS$ (1)

Draw $PU \& PV$ parallel to the axes of y and x, so that $PU = y'$ and $PV = x'$.

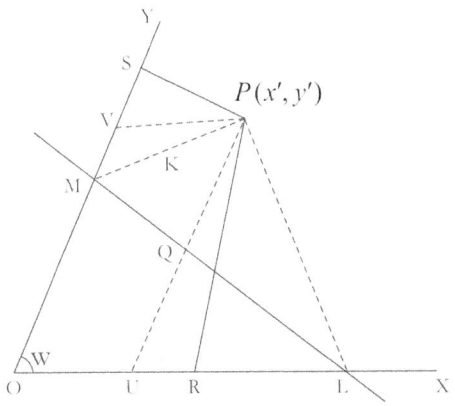

Fig. 53

Hence, $PR = PU \sin PUR = y' \sin w$

And $PS = PV \sin PVS = x' \sin w$

Also $LM = \sqrt{OL^2 + OM^2 - 2OL \cdot OM \cos w}$

$$= \sqrt{\frac{C^2}{A^2} + \frac{C^2}{B^2} - \frac{2C^2}{AB} \cos w}$$

$$LM = -C\sqrt{\frac{1}{A^2} + \frac{1}{B^2} - \frac{2\cos w}{AB}}$$

C is a negative quantity

On substituting these values in (1), we have

$$PQ - (-C)\sqrt{\frac{1}{A^2} + \frac{1}{B^2} - \frac{2\cos w}{AB}} + \frac{C^2}{AB}\sin w$$

$$= \frac{-C}{A}y'\sin w - \frac{C}{B}x'\sin w$$

$$PQ - C\sqrt{\frac{B^2 + A^2 - 2AB\cos w}{A^2 B^2}}$$

$$= C\left(\frac{y'}{A} + \frac{x'}{B} + \frac{C}{AB}\right)\sin w$$

$$PQ\frac{\sqrt{B^2 + A^2 - 2AB\cos w}}{AB}$$

$$= \frac{By' + Ax' + C}{AB}\sin w$$

$$\text{Or } PQ = \frac{Ax' + By' + C}{\sqrt{A^2 + B^2 - 2AB\cos w}}\sin w$$

Corollary:– If $w = 90°$, if the axes be rectangular, then

$$PQ = \frac{Ax' + By' + C}{\sqrt{A^2 + B^2}}$$

8. **Determine the equations to straight line passing through fixed points**

If the equation to a straight line be of the form

$$ax + by + c + \lambda(a'x + b'y + c') = 0 \tag{1}$$

Where λ is any arbitrary constant, it always passes through one fixed point, whatever be the value of λ.

The equation (1) is satisfied by the co-ordinates of the point which satisfies both of the equations.

$$ax + by + c = 0$$

And $a'x + b'y + c = 0$

By cross multiplication method, we have

$$\frac{x}{bc' - b'c} = \frac{y}{a'c - ac'} = \frac{1}{ab' - a'b}$$

$$\therefore x = \frac{bc' - b'c}{ab' - a'b}, \quad y = \frac{a'c - ac'}{ab' - a'b}$$

The point of intersection of above two lines is

$$\left(\frac{bc' - b'c}{ab' - a'b}, \frac{a'c - ac'}{ab' - a'b} \right)$$

And these co-ordinates are independent of λ.

Example 1: The axes being inclined at an angle of $30°$, obtain the equations to the straight lines which pass through the origin and are inclined at $45°$ to the straight line $x + y = 1$.

Let either of the required straight lines be $y = mx$.

The given straight line is $y = -x + 1$, so that $m' = -1$

We therefore have

$$\frac{(m - m')\sin w}{1 + (m + m')\cos w + mm'} = \tan(\pm 45°)$$

Where $m' = -1 \ \& \ w = 30°$

This equation gives

$$\frac{(m + 1)\sin 30°}{1 + (m - 1)\cos 30° - m} = \pm 1$$

If $\dfrac{(m+1)\dfrac{1}{2}}{1+(m-1)\dfrac{\sqrt{3}}{2}-m}=1$ then

$$\frac{m+1}{2+\sqrt{3}(m-1)-2m}=1$$

$$m+1=2+\sqrt{3}m-\sqrt{3}-2m$$

Or $3m-\sqrt{3}m=1-\sqrt{3}$

$$m=\frac{-\left(\sqrt{3}-1\right)}{\sqrt{3}\left(\sqrt{3}-1\right)}=\frac{-1}{\sqrt{3}}$$

$$\therefore m=\frac{-1}{\sqrt{3}}$$

If $\dfrac{m+1}{2+\left(\sqrt{3}m-\sqrt{3}\right)-2m}=-1$, then

$$m+1=-2-\sqrt{3}m+\sqrt{3}+2m$$

$$3-\sqrt{3}=m-\sqrt{3}m$$

$$\sqrt{3}\left(\sqrt{3}-1\right)=-m\left(\sqrt{3}-1\right)$$

$$m=-\sqrt{3}$$

The required equations are therefore

$$y=-\sqrt{3}x \ \& \ y=\frac{-1}{\sqrt{3}}x$$

i.e. $y+\sqrt{3}x=0$ and $\sqrt{3}\,y+x=0$

6. Equations for a Set of Straight Lines

1.   Introduction: Let us start off with the following general equation.

$$ax^2 + 2hxy + by^2 = 0 \qquad (1)$$

On multiplying it by a, we get

$$a^2x^2 + 2ahxy + aby^2 = 0$$

Or $a^2x^2 + 2ahxy + h^2y^2 - h^2y^2 + aby^2 = 0$

Or $(a^2x^2 + 2ahxy + h^2y^2) - (h^2 - ab)y^2 = 0$

Or $(ax + hy)^2 - \left(\sqrt{h^2 - ab}\right)^2 y^2 = 0$

i.e. $\left\{(ax + hy) + y\sqrt{h^2 - ab}\right\}\left\{(ax + hy) - y\sqrt{h^2 - ab}\right\} = 0$

Equation (1) represents the two straight lines whose equations are

$$ax + hy + y\sqrt{h^2 - ab} = 0 \qquad (2)$$

$$ax + hy - y\sqrt{h^2 - ab} = 0 \qquad (3)$$

Each of which passes through the origin.

(1) is satisfied by all the points which satisfy (2) & all the points which satisfy (3).

Note:– Two straight lines are real and different if $h^2 > ab$ real and coincident if $h^2 = ab$ and imaginary if $h^2 < ab$.

2.   The axes being rectangular, to find the angle between the straight lines given by the equation $ax^2 + 2hxy + by^2 = 0$ **(1)**

Let the separate equations to the lines be

$$y - m_1x = 0 \text{ and } y - m_2x = 0 \qquad (2)$$

So that (1), must be equivalent to

$$b(y - m_1x)(y - m_2x) = 0 \qquad (3)$$

Equating the co-efficient of x y and x^2 in (1) and (3), we have

$$-b(m_1 + m_2) = 2h \text{ and } bm_1m_2 = a$$

So that $m_1 + m_2 = \dfrac{-2h}{b}$ and $m_1m_2 = \dfrac{a}{b}$

If θ be the angle between the straight lines (2), we have

$$\tan\theta = \frac{m_1 - m_2}{1 + m_1m_2}$$

$$= \frac{\sqrt{(m_1 + m_2)^2 - 4m_1m_2}}{1 + m_1m_2}$$

$$= \frac{\sqrt{\dfrac{4h^2}{b^2} - 4\dfrac{a}{b}}}{1 + \dfrac{a}{b}}$$

$$= \frac{\dfrac{2\sqrt{h^2 - ab}}{b}}{\dfrac{b+a}{b}} = \frac{2\sqrt{h^2 - ab}}{a+b}$$

$$\therefore \theta = \tan^{-1}\left(\frac{2\sqrt{h^2 - ab}}{a+b}\right)$$

If the algebraic sum of the coefficients of x^2 and y^2 is zero, then we know that the straight lines are perpendicular to one another.

i.e. if $a + b = 0$ then $\tan\theta = \infty \Rightarrow \theta = 90°$

Condition for the straight lines to be coincident is

$$h^2 = ab$$

If $h^2 = ab$ then $\tan\theta = 0$

$$\Rightarrow \theta = 0$$

3. Let us now turn our attention to angular bisectors. Let is determine the equation to the straight lines bisecting the angle between the straight lines given by $ax^2 + 2hxy + by^2 = 0$.

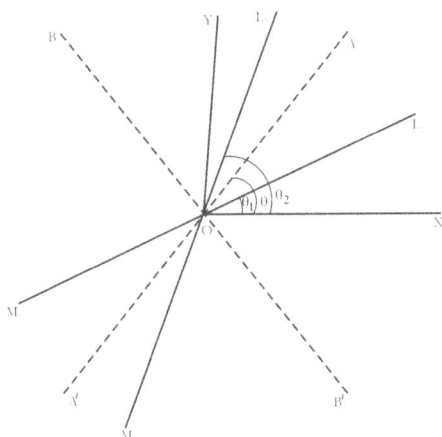

Fig.54

$ax^2 + 2hxy + by^2 = 0$ (1)

Let the equation (1), represents the two straight lines L_1OM_1 and L_2OM_2 inclined at angles θ_1 and θ_2 to the axis of x. So that the equation (1) is equivalent to $b(y - x\tan\theta_1)$
$(y - x\tan\theta_2) = 0$.

Hence $\tan\theta_1 + \tan\theta_2 = \dfrac{-2h}{b}$

And $\tan\theta_1 \tan\theta_2 = \dfrac{a}{b}$ (2)

Let OA and OB be the required bisectors.

Since, $\lfloor AOL_1 = \lfloor L_2OA$

$\therefore \lfloor AOX - \theta_1 = \theta_2 - \lfloor AOX$

$\therefore 2\lfloor AOX = \theta_1 + \theta_2$

Also $\lfloor BOX = 90° + \lfloor AOX$

$$\therefore 2\lfloor BOX = 180° + \theta_1 + \theta_2$$

Hence if θ stand for either of the angles AOX or BOX, we have

$$\tan 2\theta = \tan(\theta_1 + \theta_2) = \frac{\tan\theta_1 + \tan\theta_2}{1 - \tan\theta_1 \tan\theta_2} = -\frac{2h}{b-a}$$

By equation (2)

But, if (x, y) be the co-ordinates of any point on either of the lines OA or OB, we have

$$\tan\theta = \frac{y}{x}$$

$$\therefore \frac{-2h}{b-a} = \tan 2\theta = \frac{2\tan\theta}{1 - \tan^2\theta}$$

$$\frac{-2h}{b-a} = \frac{\dfrac{2y}{x}}{1 - \dfrac{y^2}{x^2}}$$

Or $\dfrac{-2h}{b-a} = \dfrac{2xy}{x^2 - y^2}$

Or $\dfrac{x^2 - y^2}{a-b} = \dfrac{xy}{h}$

This being a relation holding between the co-ordinates of any point on either of the bisectors is the equation to the bisectors.

Note: General Equation of the Second Degree:

$ax^2 + 2hxy + by^2 + 2gx + 2fy + c = 0$ is **called general equation of second degree.**

4. Let us now determine the condition that the general equation of the second $ax^2 + 2hxy + by^2 + 2gx + 2fh + c = 0$ **(1)** may represent two straight lines.

If we can break the left-hand members of (1) into two factors, each of the first degree, then it will represent two straight lines.

If a be not zero, multiply eq (1) by a and arrange in powers of x, it then becomes

$$a^2x^2 + 2ax(hy + g) = -aby^2 - 2afy - ac$$

On completing the squares on the left hand we have

$$a^2x^2 + 2ax(hy + g) + (hy + g)^2$$
$$= -aby^2 - 2afy - ac + (hy + g)^2$$
$$= -aby^2 - 2afy - ac + h^2y^2 + g^2 + 2ghy$$
$$= y^2(h^2 - ab) + 2y(gh - af) + g^2 - ac$$

i.e. $\{ax + (hy + g)\}^2 = y^2(h^2 - ab) + 2h(gh - af) + g^2 - ac$

$$ax + (hy + g) = \pm\sqrt{y^2(h^2 - ab) + 2y(gh - af) + g^2 - ac}$$
$$(2)$$

From (2) we cannot obtain x in terms of y, involving only terms of the first degree, unless the quantity under the radical sign be a perfect square

The condition for this is,

$$(gh - af)^2 = (h^2 - ab)(g^2 - ac)$$

i.e. $g^2h^2 - 2afgh + a^2f^2 = g^2h^2 - abg^2 - ach^2 + a^2bc$

Cancelling and dividing by a, we have the required condition

$$abc + 2fgh - af^2 - bg^2 - ch^2 = 0 \tag{3}$$

The quantity on the left hand side of eq (3) is called the discriminant of the General equation.

The general equation therefore, represents two straight lines if its discriminant be zero.

5. **Let now prove that a homogeneous equation of the nth degree represents n straight lines, real or imaginary, which all pass through the origin.**

    Let the equation be,

    $$y^n + A_1 y^{n-1}x + A_2 x^2 y^{n-2} + \cdots + A_n x^n = 0$$

    On dividing by x^n, we get

    $$\left(\frac{y}{x}\right)^n + A_1\left(\frac{y}{x}\right)^{n-1} + A_2\left(\frac{y}{x}\right)^{n-2} + \cdots + A_n = 0 \quad (1)$$

    This is an equation of the nth degree in $\frac{y}{x}$ and hence must have n roots.

    Let these roots be $m_1, m_2, m_3, \cdots m_n$. Then the equation (1) must be equivalent to the equation

    $$\left(\frac{y}{x} - m_1\right)\left(\frac{y}{x} - m_2\right)\left(\frac{y}{x} - m_3\right)\cdots\left(\frac{y}{x} - m_n\right) = 0 \quad (2)$$

    The equation (2) is satisfied by all the points which satisfy the separate equations

    $$\frac{y}{x} - m_1 = 0, \frac{y}{x} - m_2 = 0, \cdots \frac{y}{x} - m_n = 0$$

    i.e. by all the points which lie on the n straight lines

    $$y - m_1 x = 0, y - m_2 x = 0, \cdots y - m_n x = 0$$

    All of which pass through the origin.

    Example 1:–

    The equation $y^3 - 6xy^2 + 11x^2 y - 6x^3 = 0$

    Which is equivalent to $(y - x)(y - 2x)(y - 3x) = 0$

    It represent three straight lines

    $$y - x = 0, y - 2x = 0 \text{ and } y - 3x = 0$$

    All of which pass through the origin.

Solved Examples

Example 1:-- Prove that the equation $12x^2 + 7xy - 10\,y^2 + 13x + 45\,y - 35 = 0$ represents two straight lines, and find the angle between them.

Comparing given equation with $ax^2 + 2hxy + by^2 + 2gx + 2fy + c = 0$ we have $a = 12$, $h = \dfrac{7}{2}$,

$b = -10$, $g = \dfrac{13}{2}$, $f = \dfrac{45}{2}$ & $c = -35$

Hence $abc + 2\,fgh - af^2 - bg^2 - ch^2$

$$= 12 \times -10 \times -35 + 2 \times \frac{45}{2} \times \frac{13}{2} \times \frac{7}{2} - 12 \times \left(\frac{45}{2}\right)^2$$

$$-(-10) \times \left(\frac{13}{2}\right)^2 - 1(-35)\left(\frac{7}{2}\right)^2$$

$$= 4200 + \frac{4095}{4} - 6075 + \frac{1690}{4} + \frac{1715}{4}$$

$$= -1875 + \frac{7500}{4}$$

$$= 0$$

The equation represents two straight lines

Solving it for x, we have

$$x^2 + x\left(\frac{7\,y + 13}{12}\right) + \left(\frac{7\,y + 13}{24}\right)^2$$

$$= \frac{10\,y^2 - 45\,y + 35}{12} + \left(\frac{7\,y + 13}{24}\right)^2$$

$$= \left(\frac{23\,y - 43}{24}\right)^2$$

$$\therefore x + \frac{7y+13}{24} = \pm\frac{23y-43}{24}$$

i.e. $x = \dfrac{2y-7}{3}$ or $\dfrac{-5y+5}{4}$

The given equation therefore, represent the two straight lines

$$3x = 2y - 7 \text{ and } 4x = -5y + 5$$

The m's of these two lines are therefore $\dfrac{3}{2}$ and $\dfrac{-4}{5}$ and the angle between them

$$= \tan^{-1}\left\{\frac{\dfrac{3}{2} - \left(\dfrac{-4}{5}\right)}{1 + \dfrac{3}{2}\left(\dfrac{-4}{5}\right)}\right\}$$

$$= \tan^{-1}\left\{\frac{\dfrac{15+8}{10}}{\dfrac{10-12}{10}}\right\}$$

$$= \tan^{-1}\left(\frac{-23}{2}\right)$$

Example 2:– Find the value of h so that the equation $6x^2 + 2hxy + 12y^2 + 22x + 31y + 20 = 0$ may represent two straight lines.

Here $a = 6, b = 12, g = 11, f = \dfrac{31}{2}$ and $c = 20, h = h$

$$\therefore abc + 2fgh - af^2 - bg^2 - ch^2 = 0$$

$$6(12)(20) + 2\left(\frac{31}{2}\right)(11)h - 6\left(\frac{31}{2}\right)^2 - 12(11)^2 - 20h^2$$

$$= 20h^2 - 341h + \frac{2907}{2} = 0$$

i.e. $\left(b - \dfrac{17}{2}\right)(20b - 171) = 0$

Hence $b = \dfrac{17}{2}$ or $\dfrac{171}{20}$

Taking $b = \dfrac{17}{2}$, the given equation becomes

$$6x^2 + 17xy + 12y^2 + 22x + 31y + 20 = 0$$

i.e. $(2x + 3y + 4)(3x + 4y + 5) = 0$

Taking $b = \dfrac{171}{20}$, the equation is

$$20x^2 + 57xy + 40y^2 + \frac{220}{3}x + \frac{310}{3}y + \frac{200}{3} = 0$$

i.e. $\left(4x + 5y + \dfrac{20}{3}\right)(5x + 8y + 10) = 0$

Example 3: To find the equation to the two straight lines joining the origin to the points in which the straight line $lx + my = n$ meets the locus whose equation is $ax^2 + 2hxy + by^2 + 2gx + 2fy + c = 0$

$lx + my = n$ (1)

$ax^2 + 2hxy + by^2 + 2gx + 2fy + c = 0$ (2)

From (1) we have $\dfrac{lx + my}{n} = 1$ \hfill (3)

The co-ordinates of the point in which the straight line meets the locus satisfy both equation (2) & ex (3), and hence satisfy the equation

$$ax^2 + 2hxy + by^2 + 2(gx + fy)\left(\frac{lx + my}{n}\right) + c\left(\frac{lx + my}{n}\right)^2 = 0$$

(4)

Hence (4), represents some locus which passes through the intersection of (2) & (3).

But, since the equation (4), is homogeneous and of the second degree, it represents two straight lines passing through the origin.

Example 4: What is represented by the equation $(x^2 - a^2)^2 + (y^2 - b^2)^2 = 0$

The only real points on the locus are those that satisfy both of the equations,

$$x^2 - a^2 = 0 \text{ and } y^2 - b^2 = 0$$
$$x = \pm a \quad y = \pm b$$

The points represented are therefore

$$(a,b) \ (a,-b) \ (-a,b) \ \& \ (-a,-b)$$

Example 5: Obtain the condition that one of the straight lines given by the equation $ax^2 + 2hxy + by^2 = 0$ may coincide with one of those given by the equation $a'x^2 + 2h'xy + b'y^2 = 0$ (2)

$$ax^2 + 2hxy + by^2 = 0 \ (1)$$
$$a'x^2 + 2h'xy + b'y^2 = 0 \ (2)$$

Let the equation to the common straight line be

$$y - m_1 x = 0 \tag{3}$$

The quantity $y - m_1 x$ must therefore be a factor of the left-hand of both (1) & (2), and therefore, the value $y = m_1 x$ must satisfy both (1) & (2).

We therefore, have

$$bm_1^2 + 2hm_1 + a = 0 \tag{4}$$

And $b'm_1^2 + 2h'm_1 + a' = 0$ (5)

Solving (4) & (5), we have.

$$\frac{m_1^2}{2(ha' - h'a)} = \frac{m_1}{ab' - a'b} = \frac{1}{2(bh' - b'h)}$$

$$\therefore \frac{ba' - b'a}{bb' - b'b} = m_1{}^2 = \left\{ \frac{ab' - a'b}{2(bb' - b'b)} \right\}^2$$

So that, we must have

$$(ab' - a'b)^2 = 4(ba' - b'a)(bb' - b'b)$$

Example 6: Prove that two of the straight lines represented by the equation $ax^3 + bx^2 y + cxy^2 + dy^3 = 0$ (1) will be at right angles if $a^2 + ac + bd + d^2 = 0$.

Let the separate equations to the three lines be

$$y - m_1 x = 0 , \ y - m_2 x = 0 \text{ and } y - m_3 x = 0$$

So that the equation (1) must be equivalent to

$$d(y - m_1 x)(y - m_2 x)(y - m_3 x) = 0$$

And therefore $m_1 + m_2 + m_3 = \dfrac{-c}{d}$ \hfill (2)

$$m_1 m_2 + m_2 m_3 + m_3 m_1 = \frac{b}{d} \hfill (3)$$

$$m_1 m_2 m_3 = \frac{-a}{d} \hfill (4)$$

If the first two of these straight lines be at right angles we have, in addition

$$m_1 m_2 = -1 \hfill (5)$$

From (4) & (5), we have

$$m_3 = \frac{a}{d}$$

And therefore from (2)

$$m_1 + m_2 = \frac{-c}{d} - \frac{a}{d} = -\frac{c + a}{d}$$

The equation (3), then becomes

$$\frac{a}{d}\left(-\frac{c+a}{d}\right)-1=\frac{b}{d}$$

$$-ac-a^2-d^2=bd$$

Or $a^2+ac+d^2+bd=0$

7. Transformation of Coordinates

While dealing with a coordinate system, it is desirable for us to alter the origin and its axes by doing one of three things:

1. altering the origin without alternation of the direction of the axes

2. altering the directions of the axes and keeping the origin unchanged,

3. altering the origin and also the directions of the axes.

These manipulations constitute the focus area for our discussion on Transformation of co-ordinates.

1. **Let us understand what it takes to alter the origin of co-ordinates without altering the directions of the axes.**

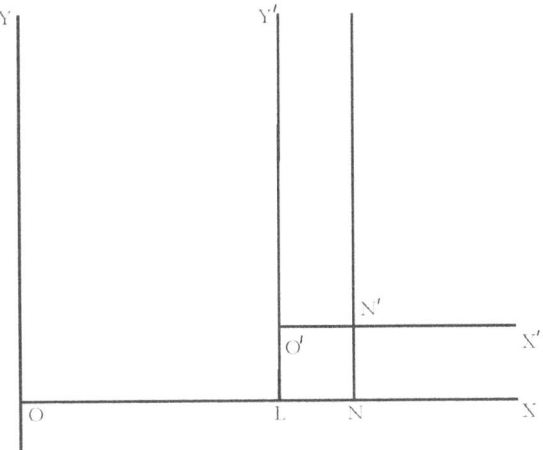

Fig. 55

Let OX and OY be the original axes and let the new axes, parallel to the original be $O'X'$ and $O'Y'$.

Let the co-ordinates of the new origin O', referred to the original axes be h and k, so that, if $O'L$ be $\perp$ to OX, we have

$$OL = h \text{ and } LO' = K$$

Let P be any point in the plane of the paper, and let it co-ordinates referred to the original axes be x and y and referred to the new axes let them be x' and y'.

Draw $PN \perp$ to OX to meet $O'X'$ in N' then, $ON = x$, $NP = y$, $O'N' = x'$ and $N'P = y'$

We therefore, have

$$x = ON = OL + O'N' = h + x'$$
$$y = NP = LO' + N'P = K + y'$$

The origin is therefore, transferred to the point (h, k) when we substitute for the co-ordinates x and y the quantities $x' + h$ and $y' + k$.

Note: The above result is true, if axes be oblique or rectangular.

2. **Let us now change the direction of the axes of co-ordinates, without changing the origin, both systems of co-ordinates being rectangular.**

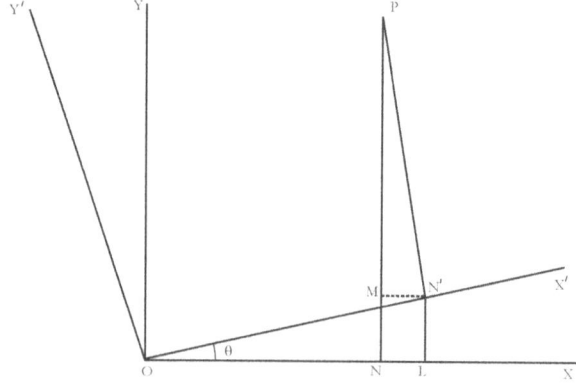

Fig. 56

Let OX and OY be the original system of axes and OX' and OY' $\lfloor XOX'$,through which the axes are turned be called θ .

Take any point P in the plane of the paper.

Draw PN and $PN' \perp$ to OX and OX' , and also $N'L \, \& \, N'M \perp$ to OX and PN.

If the co-ordinates of P, referred to the original axes be x and y, and referred to the new axes, be x' and y'

We have $ON = x, \;\; NP = y, ON' = x'$ and $N'P = y'$

The angle

$$MPN' = 90° - \lfloor MN'P = \lfloor MN'O = \lfloor XOX' = \theta$$

We then have

$$x = ON = OL - MN' = ON'\cos\theta - N'P\sin\theta$$
$$= x'\cos\theta - y'\sin\theta \tag{1}$$

And $y = NP = LN' + MP = ON'\sin\theta + N'P\cos\theta$

$$= x'\sin\theta + y'\cos\theta \tag{2}$$

If therefore, in any equation, we wish to turn the axes, being rectangular, through an angle θ , we must substitute $x'\cos\theta - y'\sin\theta$ and $x'\sin\theta + y'\cos\theta$,for x and y.

Note:–

1. When we have both to change the origin, and also the direction of the axes, the transformation is clearly obtained by combining the result of the previous articles.

If the origin is to be transformed to the point (h, k) and the axes to be turned through an angle θ , we have to substitute $h + x'\cos\theta - y'\sin\theta$ and $k + x'\sin\theta + y'\cos\theta$ for x and y respectively.

3. **Let us now change from one set of axes, inclined at angle w, to another set, inclined at an angle w' , the origin remaining unaltered.**

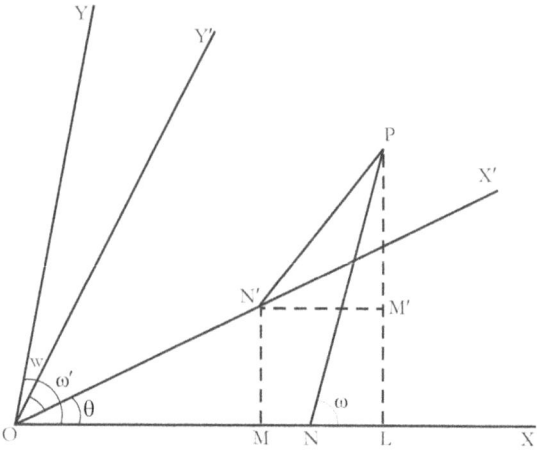

Fig. 57

Let OX & OY be the original axes, OX' & OY' the new axes, and let the angle XOX' be θ. $\lfloor X'OY' = w'$

Take any point P in the plane of the paper.

Draw PN and PN' parallel to OY and OY' to meet OX & OX' respectively in N and N', PL $\perp$ to OX, and $N'M$ & $N'M' \perp$ to OL and LP.

Now, $\lfloor PNL = \lfloor YOX = w$

And $\lfloor PN'M' = \lfloor Y'OM = w' + \theta$.

Hence if $ON = x$, $NP = y$, $ON' = x'$ and $N'P = y'$.

We have $y \sin w = NP \sin w = LP = MN' + M'P$

$$= ON' \sin\theta + N'P \sin(w' + \theta)$$

So that $y \sin w = x' \sin\theta + y' \sin(w' + \theta)$ \hfill (1)

Also $x + y \cos w = ON + NL = OL = OM + N'M'$

$$= x' \cos\theta + y' \cos(w' + \theta)$$ \hfill (2)

Multiplying (2), by $\sin w$, (1) by $\cos w$, and subtracting, we have

$$x \sin w = x' \sin(w - \theta) + y' \sin(w - w' - \theta) \qquad (3)$$

The equations (1) & (3), give the proper substitutions for the change of axes in the general case

Special Cases:

Case 1: Suppose we transfer our axes from a rectangular pair to one inclined at an angle w'. In this case w is $90°$, and the formulae becomes

$$x = x' \cos\theta + y' \cos(w' + \theta)$$

And $y = x' \sin\theta + y' \sin(w' + \theta)$

Case 2: Suppose the transference is to be from oblique axes, inclined at w, to rectangular axes, In this case w' is $90°$. And formulae becomes

$$x \sin w = x' \sin(w - \theta) - y' \cos(w - \theta)$$

And $y \sin w = x' \sin\theta + y' \cos\theta$.

4. **We will determine the angle through which the axes must be turned so that the expression $ax^2 + 2hxy + by^2$ may become an expression in which there is no term involving $x'y'$.**

If $h' = 0$ then from (3) of last article
$$0 = 2h \cos 2\theta - (a - b) \sin 2\theta$$

Or $2h \cos 2\theta = (a - b) \sin 2\theta$

Or $\tan 2\theta = \dfrac{2h}{a - b}$

The required angle is therefore

$$2\theta = \tan^{-1}\left(\frac{2h}{a - b}\right)$$

Or $\theta = \dfrac{1}{2}\tan^{-1}\left(\dfrac{2h}{a-b}\right)$

If by any change of axes, without change of origin, the quantity $ax^2 + 2hxy + by^2$ becomes $a'x^2 + 2h'xy + b'y^2$, then

$$\frac{a+b-2h\cos w}{\sin^2 w} = \frac{a'+b'-2h'\cos w'}{\sin^2 w'}$$

And $\dfrac{ab-h^2}{\sin^2 w} = \dfrac{a'b'-h'^2}{\sin^2 w'}$

W and w' are the angles between the original and final pairs of axes.

Let the co-ordinates of any point P, referred to the original axes be x & y and referred to the final axes, let them be x' & y'.

The square of the distance between P and the origin is $x^2 + 2xy\cos w + y^2$, referred to the original axes and $x'^2 + 2x'y'\cos w' + y'^2$, referred to the final axes.

We therefore, always have

$$x^2 + 2xy\cos w + y^2 = x'^2 + 2x'y'\cos w' + y'^2 \qquad (1)$$

Also, by supposition, we have

$$ax^2 + 2hxy + by^2 = a'x'^2 + 2h'x'y' + b'y'^2 \qquad (2)$$

Multiplying (1) by λ and adding it to (2), we therefore, have

$$x^2(a+\lambda) + 2xy(h+\lambda\cos w) + y^2(b+\lambda)$$
$$= x'^2(a'+\lambda) + 2x'y'(h'+\lambda\cos w') + y'^2(b'+\lambda) \qquad (3)$$

If then any value of λ makes the LHS of (3), a perfect square, the same value must make the RHS also a perfect square.

But the values of λ which make the LHS a perfect square are given by the condition

$$(h+\lambda\cos w)^2 = (a+\lambda)(b+\lambda)$$

i.e. $h^2(1-\cos^2 w) + \lambda(a+b-2h\cos w) + ab - h^2 = 0$

i.e. by $\lambda^2 + \lambda \dfrac{a+b-2h\cos w}{\sin^2 w} + \dfrac{ab-h^2}{\sin^2 w} = 0$ (4)

In a similar manner the values of λ which make the RHS of (3), a perfect square are given by the equation

$$\lambda^2 + \lambda \dfrac{a'+b'-2h'\cos w'}{\sin^2 w} + \dfrac{a'b'-h'^2}{\sin^2 w} = 0 \quad (5)$$

Since the values of λ given by eq (4), are the same as the values of λ given by (5), the two equations (4) & (5) must be the same.

Hence, we have

$$\frac{a+b-2h\cos w}{\sin^2 w} = \frac{a'+b'-2h'\cos w'}{\sin^2 w'}$$

And $\dfrac{ab-h^2}{\sin^2 w} = \dfrac{a'b'-h'^2}{\sin^2 w'}$

Solved Examples

Example 1: Transform to parallel axes through the point $(-2,3)$ the equation $2x^2 + 4xy + 5y^2 - 4x - 22y + 7 = 0$.

We substitute $x = x' - 2$ and $y = y' + 3$, and the equation becomes,

$$2(x'-2)^2 + 4(x'-2)(y'+3) + 5(y'+3)^2$$
$$-4(x'-2) - 22(y'+3) + 7 = 0$$
$$2(x'^2 + 4 - 4x') + 4x'y' + 12x' - 8y' - 24 + 5(y'^2 + 9 + 6y')$$
$$-4x' + 8 - 22y' - 66 + 7 = 0$$
$$2x'^2 + 5y'^2 + 4x'y' - 22 = 0$$

Example 2: Transform to axes inclined at $30°$ to the original axes the equation $x^2 + 2\sqrt{3}xy - y^2 = 2a^2$.

For x and y we have to substitute

$$x' \cos 30° - y' \sin 30° \text{ and } x' \sin 30° + y' \cos 30°$$

i.e. $x' \cdot \dfrac{\sqrt{3}}{2} - y' \cdot \dfrac{1}{2} \,\&\, x' \dfrac{1}{2} + y' \dfrac{\sqrt{3}}{2}$

i.e. $\dfrac{x'\sqrt{3}-1}{2} \text{ and } \dfrac{x'+y'\sqrt{3}}{2}$.

The equation then becomes

$$\left(x'\sqrt{3}-y'\right)^2 + 2\sqrt{3}\left(x'\sqrt{3}-y'\right)\left(x'+y'\sqrt{3}\right)$$

$$-\left(x'+y'\sqrt{3}\right)^2 = 8a^2$$

$$3x'^2 + y'^2 - 2\sqrt{3}x'y' + 6x'^2 + 6x'y' - 2\sqrt{3}x'y'$$

$$-6y'^2 - \left(x'^2 + 3y'^2 + 2\sqrt{3}x'y'\right) = 8a^2$$

i.e. $x'^2 - y'^2 = a^2$

Example 3: Transform the equation $\dfrac{x^2}{a^2} - \dfrac{y^2}{b^2} = 1$, from rectangular axes to axes inclined at an angle 2α, the new axis of x being inclined at an angle $-\alpha$ to the old axes and $\sin\alpha$ being equal to $\dfrac{-b}{\sqrt{a^2+b^2}}$.

Here $\theta = -\alpha$ and $w' = 2\alpha$, so that the formulae of transformation (1) becomes,

$$x = (x'+y')\cos\alpha \text{ and } y = (y'-x')\sin\alpha$$

Since, $\sin\alpha = \dfrac{b}{\sqrt{a^2+b^2}}$, we have $\cos\alpha = \dfrac{a}{\sqrt{a^2+b^2}}$ and hence, the given equation becomes

$$\dfrac{(x'+y')^2}{a^2+b^2} - \dfrac{(y'-x')^2}{a^2+b^2} = 1$$

i.e. $x'y' = \dfrac{1}{4}(a^2 + b^2)$

The degree of an equation is unchanged by any transformation of co-ordinates:

The most general formulae of transformation are

$$x = h + x'\frac{\sin(w - \theta)}{\sin w} + y'\frac{\sin(w' + \theta)}{\sin w}$$

And $y = k + x'\dfrac{\sin\theta}{\sin w} + y'\dfrac{\sin(w' + \theta)}{\sin w}$

For x & y we have therefore, to substitute expression in x' and y' of the first degree, so that by this substitution the degree of the equation cannot be raised.

If by any change of axes, without change of origin, the quantity $ax^2 + 2hxy + by^2$ become $a^2x'^2 + 2h'x'y' + b'y'^2$ the axes in each case being rectangular, to prove that $a + b = a' + b'$ and $ab - h^2 = a'b' - h'^2$.

The new axis of x being inclined at an angle θ to the old axis, we have to substitute.

$x'\cos\theta - y'\sin\theta$ and $x'\sin\theta + y'\cos\theta$ for x & y respectively.

Hence $ax^2 + 2hxy + by^2$

$$= a(x'\cos\theta - y'\sin\theta)^2 + 2h(x'\cos\theta - y'\sin\theta)$$

$$(x'\sin\theta + y'\cos\theta) + b(x'\sin\theta + y'\cos\theta)^2$$

$$= x'^8[a\cos^2\theta + 2h\cos\theta\sin\theta + b\sin^2\theta]$$

$$+ 2x'y'[-a\cos\theta\sin\theta + h(\cos^2\theta - \sin^2\theta) + b\cos\theta\sin\theta]$$

$$+ y'^2[a\sin^2\theta - 2h\cos\theta\sin\theta + b\cos^2\theta]$$

We then have

$$a' = a\cos^2\theta + 2h\cos\theta\sin\theta + b\sin^2\theta$$

$$= \frac{1}{2}[(a+b)+(a-b)\cos 2\theta + 2h\sin 2\theta]$$

$$(1)$$

$$b' = a\sin^2\theta - 2h\cos\theta\sin\theta + b\cos^2\theta$$

$$= \frac{1}{2}[(a+b)-(a-b)\cos 2\theta - 2h\sin 2\theta] \qquad (2)$$

And $h' = -a\cos\theta\sin\theta + b(\cos^2\theta - \sin^2\theta) + b\cos\theta\sin\theta$

$$= \frac{1}{2}[2h\cos 2\theta - (a-b)\sin 2\theta] \qquad (3)$$

By adding (1) & (2), we have

$$a' + b' = \frac{1}{2}[(a+b)+(a-b)\cos 2\theta + 2h\sin 2\theta$$

$$+(a+b)-(a-b)\cos 2\theta - 2h\sin 2\theta]$$

$$= \frac{1}{2}(2a+2b)$$

$$\therefore a' + b' = a + b$$

Also by multiplying (1) & (2), we have

$$a'b' = \frac{1}{4}\left\{(a+b)^2 - [(a-b)\cos 2\theta + 2h\sin 2\theta]^2\right\}$$

Hence $4a'b' = (a+b)^2 - \left\{(a-b)\cos 2\theta + 2h\sin 2\theta\right\}^2$

Hence $4a'b' - 4h'^2 = (a+b)^2 - [\{2h\sin 2\theta + (a-b)\cos 2\theta\}^2$

$$+\{2h\cos 2\theta - (a-b)\sin 2\theta\}^2]$$

$$= (a+b)^2 - [(a-b)^2 + 4h^2]$$

$$= a^2 + b^2 + 2ab - a^2 - b^2 + 2ab - 4h^2$$

$$4a'b' - 4h'^2 = 4ab - 4h^2$$

So $a'b' - h'^2 = ab - h^2$

8. The Circle

1. **Definition:**A circle is the locus of a point which moves so that its distance from a fixed point, called the centre, is equal to a given distance. The given distance is called the radius of the circle.

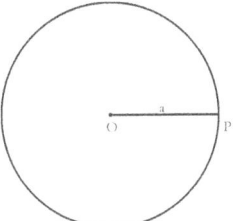

Fig.58

O is called centre and $OP = a$ is known as the radius of the circle.

2. **Let us start our exploration with the following problem. Let us determine the equation to a circle, the axes of co-ordinates being two straight lines through it centre at right angle.**

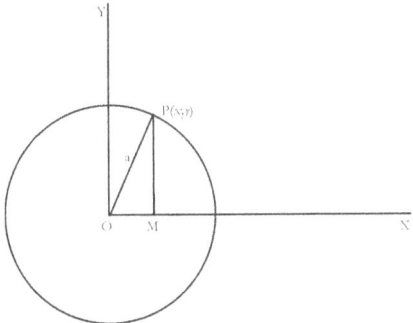

Fig.59

Let O be the centre of the circle and let a be its radius. Let OX and OY be the axes of co-ordinates. Let P be any point on the circumference of the circle, and let its co-ordinates be x and y.

Draw PM $\perp$ to OX and join OP

Then $OM^2 + MP^2 = a^2$

$$x^2 + y^2 = a^2$$

3. **Let us now generalize this step further. Let us determine the equation to a circle referred to any rectangle axes.**

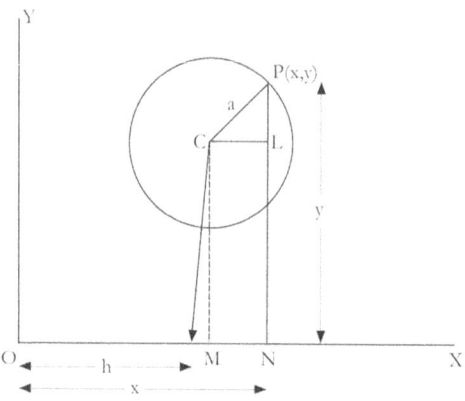

Fig. 60

Let OX and OY be the two rectangular axes.

Let C be the centre of the circle and a its radius

Take any point P on the circumference and draw ⊥ CM and PN upon OX; let P be the point (x, y)

Draw CL ⊥ to NP. Let the co-ordinates of C be h & k these are supposed to be known

We have $OM = h$, $CM = k$

$$ON = x \quad NP = y$$

$$CL = MN = ON - OM$$

$$= x - h$$

And

$$LP = PN - LN = PN - CM = y - k$$

From the triangle PCL

$$CP^2 = CL^2 + LP^2$$

108

$$a^2 = (x - h)^2 + (y - k)^2$$

$$\therefore (x - h)^2 + (y - k)^2 = a^2$$

Represents equation of circle whose centre is (h, k) and radius is equal to a.

a. **Scenario 1:**

Let the origin O be on the circle so that, in this case

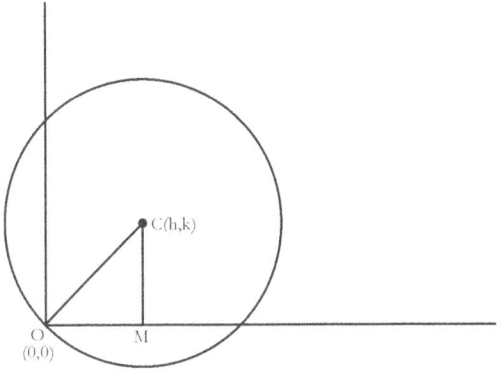

Fig. 61

$$OM^2 + MC^2 = a^2$$

$$h^2 + k^2 = a^2$$

The equation $(x - h)^2 + (y - k)^2 = a^2$ becomes

$$(x - h)^2 + (y - k)^2 = h^2 + k^2$$

Or $x^2 + h^2 - 2hx + y^2 + k^2 - 2yk = h^2 + k^2$

Or $x^2 + y^2 - 2hx - 2ky = 0$

b. **Scenario 2:**

Let the origin be not on the curve, but let the centre lie on the axis of x. In this case k = 0, and the equation becomes $(x - h)^2 + y^2 = a^2$

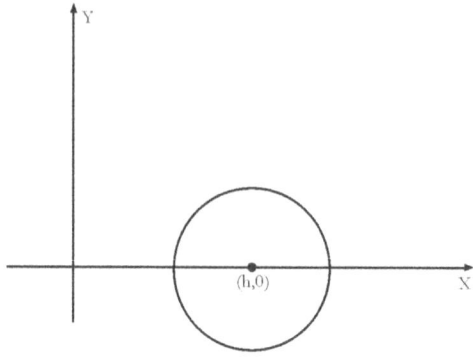

Fig.62

c. Scenario 3:

Let the origin be on the curve and let the axis of x be a diameter. We now have k = 0 and a = h, so that the equation becomes

$$x^2 + y^2 - 2hx = 0$$

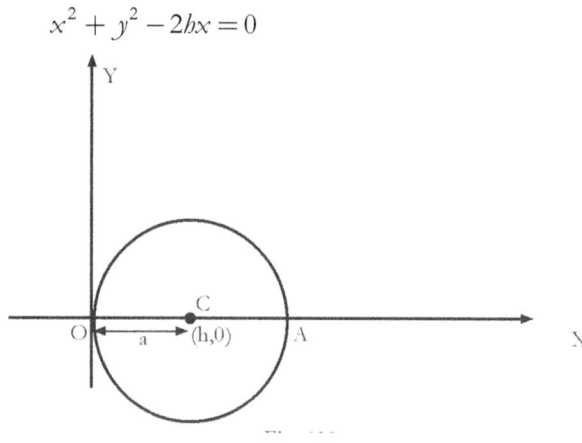

Fig. 63

d. Scenario 4:

By taking O to C , and thus making both h and k zero we have the case $x^2 + y^2 = a^2$ $(\therefore h = 0 = k)$

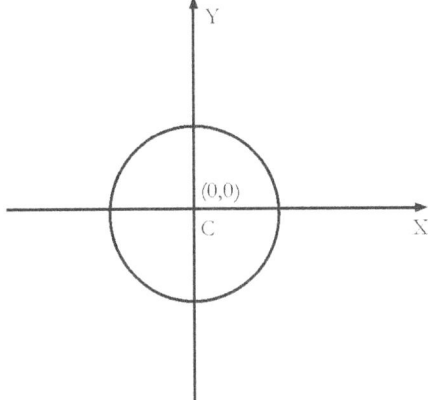

Fig. 64

e. Scenario 5:

The circle will touch the axis of x if MC is equal to the radius, if $k = a$

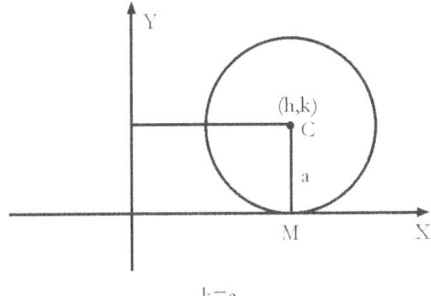

k=a

Fig. 65

$$\therefore (x - h)^2 + (y - k)^2 = a^2 \text{ becomes}$$

$$(x - h)^2 + (y - a)^2 = a^2$$

$$\text{Or } x^2 + h^2 - 2xh + y^2 + a^2 - 2ya = a^2$$

$$\text{Or } x^2 + y^2 - 2hx - 2ky + h^2 = 0 \ (\therefore k = a)$$

Similarly, one touching the axis of y is $h = a$

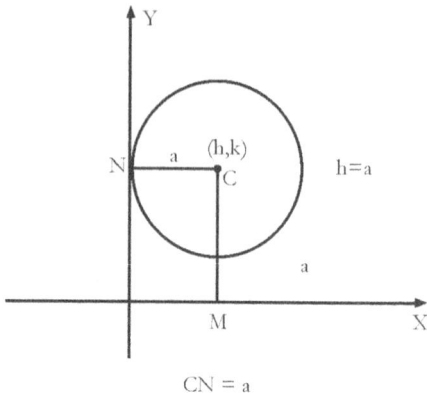

CN = a

Fig. 66

$$\therefore (x - a)^2 + (y - k)^2 = a^2$$

$$\text{Or } x^2 + a^2 - 2xa + y^2 + k^2 - 2yk = a^2$$

$$\text{Or } x^2 + y^2 - 2hx - 2ky + k^2 = 0$$

4. **Now we will show that** $x^2 + y^2 + 2gx + 2fy + c = 0$ **always represents a circle for all values of g, f and c, andwe will find its centre and radius, with the assumption that the axes are rectangular.**

Consider $x^2 + y^2 + 2gx + 2fy + c = 0$ (1)

$$x^2 + y^2 + 2gx + 2fy = -c$$

Add both sides $g^2 + f^2$, we get

$$x^2 + 2gx + g^2 + y^2 + 2fy + f^2 = g^2 + f^2 - c$$

$$\text{Or} (x + g)^2 + (y + f)^2 = g^2 + f^2 - c$$

Compare $[x - (-g)]^2 + [y - (-f)]^2 = g^2 + f^2 - c$ with

$$(x - h)^2 + (y - k)^2 = a^2$$

We have $h = -g, k = -f$ & $a^2 = g^2 + f^2 - c$

Hence (1) represents a circle whose centre $(h, k) = (-g, -f)$ and a radius $a = \sqrt{g^2 + g^2 - c}$

Note:

If $g^2 + f^2 > c$, the radius of this circle is real

If $g^2 + f^2 = c$, the radius vanishes, circle becomes point circle

If $g^2 + f^2 < c$, the radius of the circle is imaginary

5. **Observation:The general equation of the second degree in rectangular co-ordinates represents a circle, if the coefficients of x^2 and y^2 be the same and if the coefficient of x y be zero.**

6. **Let us derive the equation to a circle given the end points of its diameter. In other words, the line joining points (x_1, y_1) & (x_2, y_2) forms the diameter of the circle.**

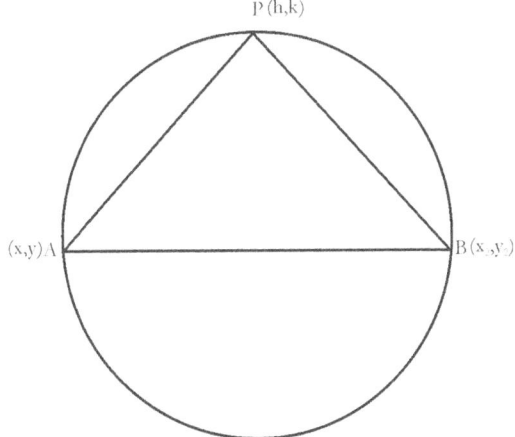

Fig. 67

Let A be the point (x_1, y_1) and B be the point (x_2, y_2) and let the co-ordinates of any point P on the circle be h and k.

The equation to AP is $A(x_1, y_1)$ $\begin{matrix} P(h,k) \\ x_2\, y_2 \end{matrix}$

$$y - y_1 = \frac{y_2 - y_1}{x_2 - x_1}(x - x_1)$$

$$\Rightarrow y - y_1 = \frac{k - y_1}{h - x_1}(x - x_1) \qquad (1)$$

The equation to BP is $\begin{matrix} B(x_2, y_2)\ P(h,k) \\ x_1\ y_1 \qquad x_2\, y_2 \end{matrix}$

$$y - y_2 = \frac{k - y_2}{h - x_2}(x - x_2) \qquad (2)$$

But, since APB is a semicircle, the angle APB is a right angle, and hence the straight lines (1) & (2) are at right angles

Product of their slopes should be -1

i.e. $\dfrac{k - y_1}{h - x_1} \cdot \dfrac{k - y_2}{h - x_2} = -1$

$Or\, (k - y_1)(k - y_2) = -(h - x_1)(h - x_2)$

$Or\, (h - x_1)(h - x_2) + (k - y_1)(k - y_2) = 0$

But this is the condition that the point (h, k) may lie on the curve whose equation is

$$(x - x_1)(x - x_2) + (y - y_1)(y - y_2) = 0$$

This is the required condition

7. **Let us determine the intercept made on the axes by the circle whose equation is** $ax^2 + ay^2 + 2gx + 2fy + c = 0$

$ax^2 + ay^2 + 2gx + 2fy + c = 0 \ (1)$

The abscissas of the points where the circle (1) meets the axis of x i.e $y = 0$, are given by the equation

$$ax^2 + 2gx + c = 0 \qquad (2)$$

The roots of this equation being x_1 and x_2, we have

$$x_1 + x_2 = \frac{-2g}{a}$$

$$\text{And } x_1 x_2 = \frac{c}{a}$$

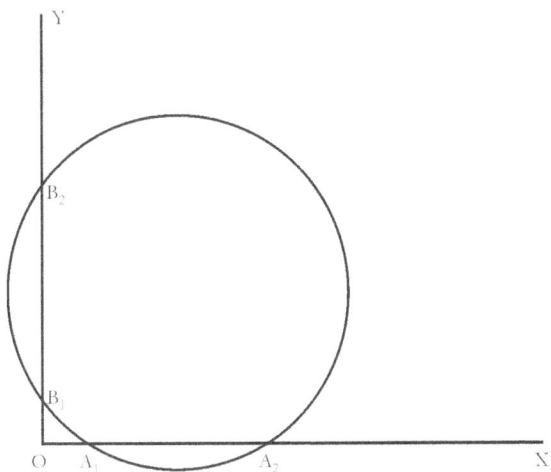

Fig. 68

Hence $A_1 A_2 = x_2 - x_1 = \sqrt{(x_1 + x_2)^2 - 4x_1 x_2}$

$$= \sqrt{\frac{4g^2}{a^2} - \frac{4c}{a}}$$

$$= \frac{2\sqrt{g^2 - ac}}{a}$$

Again, the roots of the equation (2) are both imaginary if $g^2 < ac$. In this case the circle does not meet the axis of x in real points i.e. geometrically it does not meet the axis of x at all

The circle will touch the axis of x is the intercepts $A_1 A_2$ be just zero, i.e. if $g^2 = ac$

It will meet the axis of x in two points lying on opposite sides of the origin 0 is the two roots of the equation (2) are of opposite signs i.e. if c be negative

8. **Tangent: Any straight line touches the circle at only one point is called tangent of circle at that point.**

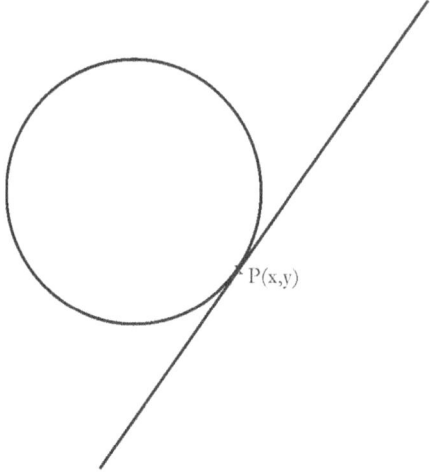

P(x,y)

Fig. 69

9. **Let us determine the equation of tangent at any point (x', y') of the circle $x^2 + y^2 = a^2$**

Let the point P be the point (x', y')

The equation to any straight line passing through P is

$$y - y' = m(x - x')$$ (1)

Also, the equation to OP is

$$y - 0 = \frac{y'}{x'}(x - 0)$$

$$y = \frac{y'}{x'}x$$

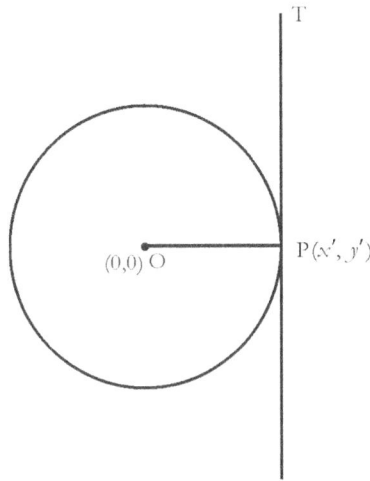

Fig.70

(2)

The straight lines (1) & (2) are at right angles then line (1) is a tangent, if

$$m\frac{y'}{y'} = -1$$

If $m = \frac{-x'}{y'}$

Substituting this value of m in (1), the equation of the tangent at (x', y') is

$$y - y' = -\frac{x'}{y'}(x - x')$$

Or $yy' - (y')^2 = -xx' + (x')^2$

Or $xx' + yy' = x'^2 + y'^2$
(3)

But, since (x', y') lies on the circle, we have $x'^2 + y'^2 = a^2$ and the required equation is than

$$xx' + yy' = a^2$$

10. **Let us now obtain the equation of the tangent at any point** (x', y') **lying on the circle** $x^2 + y^2 + 2gx + 2fy + x = 0$ **centre is** $(-g, -f)$ **and radius** $= \sqrt{g^2 + f^2 - c}$

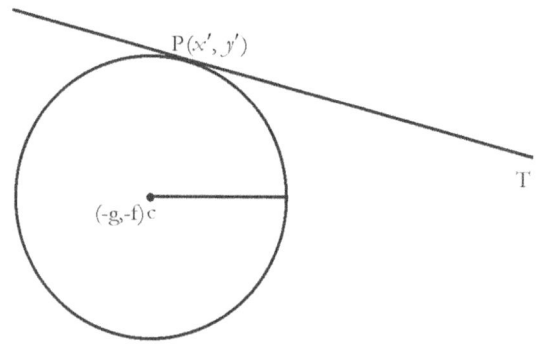

Fig.71

Let PT be the tangent to the circle and P be (x', y')

By data $P(x', y')$ lie on the circle
(1)

$$\therefore x'2 + y'^2 + 2gx' + 2fy' + c = 0$$
(2)

Clearly $CP \perp$ to PT

(slope of CP) (slope of PT) $= -1$

Slope of $PT = \dfrac{-1}{\text{slope of CP}}$

Slope of $CP = \dfrac{y' + f}{x' + g}$

Slope of $PT = -\dfrac{1}{\dfrac{y' + f}{x' + g}} = \dfrac{-(x' + g)}{y' + f}$

Also the tangent passes through the point (x', y'). Thus the equation of the tangent is

$$y - y' = \left(\dfrac{x' + g}{y' + f}\right)(x - x')$$

$\text{Or}\, (y - y')(y' + f) = -(x' + g)(x - x')$

$$yy' + yf - y'^2 - y'f = -xx' + x'^2 - gx + gx'$$

$\text{Or}\, xx' + yy' + gx + fy = x'^2 + y'^2 + gx' + fy'$

Add both sides $gx' + fy' + c$, we get

$$xx' + yy' + gx + gx' + fy + fy' + c$$

$$= x'^2 + y'^2 + 2gx' + 2fy' + c$$

$$xx' + yy' + g(x + x') + f(y + y') + c = 0$$

$(\because \text{from (2)})$

Is the required equation

11. **Observation:** The equation to the tangent at (x', y') is therefore obtained from that of the circle itself by substituting xx' for x^2, yy' for y^2, $x + x'$ for $2x$ and $y + y'$ for $2y$.

Points of intersection, in general of the straight line:

$$y = mx + c \qquad\qquad (1)$$

With the circle $x^2 + y^2 = a^2$ $\qquad\qquad (2)$

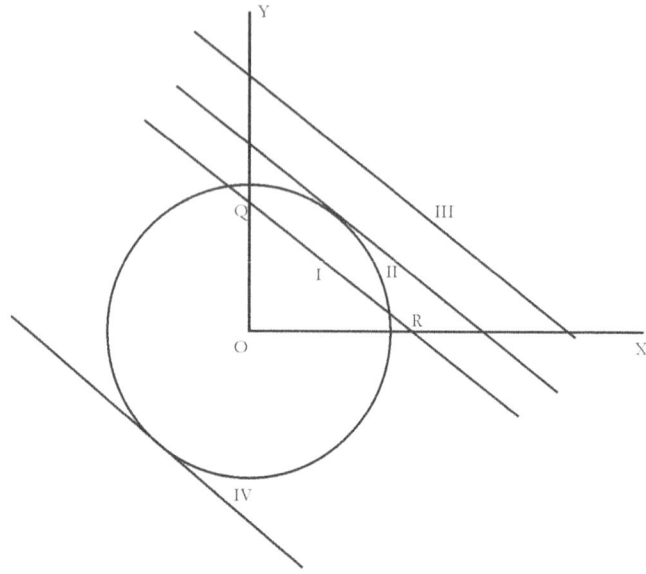

Fig.72

The co-ordinates of the points in which the straight line (1) meets (2) satisfy both equations (1) and (2)

If therefore, we solve them as simultaneous equations we shall obtain the co-ordinates of the common point or points

Substituting for y from (1) in (2), the abscissas of the required points are given by the equation

$$x^2 + (mx + c)^2 = a^2$$

i.e. $x^2 + m^2x^2 + c^2 + 2mxc = a^2$

Or $x^2(1 + m^2) + 2mcx + c^2 - a^2 = 0$ (3)

The roots of this equation are real, coincident or imaginary, according as

$$(2mc)^2 - 4(1 + m^2)(c^2 - a^2) \text{ is positive, zero, or negative}$$

i.e. according as $a^2(1 + m^2) - c^2$ is +ve, zero, or −ve

i.e. according as c^2 is $<=$ or $> a^2(1 + m^2)$

In the figure the lines marked i, ii, and iii are all parallel; their equations all have the same 'm'

The straight line 1 corresponds to a value of c^2 which is $< a^2(1+m^2)$ and it meets the circle in two real points.

The straight line 111 which corresponds to a value of $c^2 > a^2(1+m^2)$, does meet the circle at all, or rather, this is better expressed by saying that it meets the circle in imaginary points.

The straight line 11 corresponds to a value of c^2, which is equal to $a^2(1+m^2)$, and meets the curve in two coincident points i.e. is a tangent.

We can now obtain the length of the chord intercepted by the circle on the straight line (1). For if x_1 and x_2 be the roots of the equation (3), we have

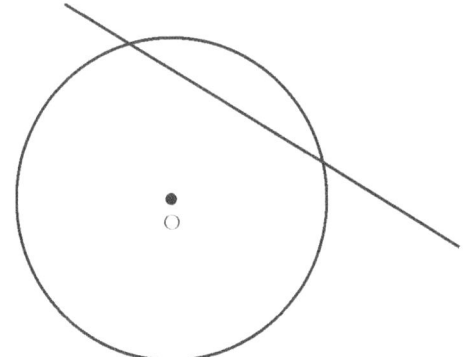

Fig.73

$$x_1 + x_2 = \frac{-2mc}{1+m^2} \text{ and } x_1 x_2 = \frac{c^2 - a^2}{1+m^2}$$

Hence

$$x_1 - x_2 = \sqrt{(x_1 + x_2)^2 - 4x_1 x_2}$$

$$= \sqrt{\frac{4m^2 c^2}{(1+m^2)^2} - \frac{4(x^2 - a^2)}{1+m^2}}$$

$$= \frac{2}{1+m^2}\sqrt{m^2c^2-(c^2-a^2)(1+m^2)}$$

$$= \frac{2}{1+m^2}\sqrt{m^2c^2-c^2-c^2m^2+a^2+a^2m^2}$$

$$= \frac{2}{1+m^2}\sqrt{a^2(1+m^2)-c^2}$$

If y_1 and y_2 be the ordinates of Q and R we have, since these points are on, (1)

$$y_1-y_2=(mx_1+c)-(mx_2+c)$$
$$= m(x_1-x_2)$$

Hence $QR = \sqrt{(y_1-y_2)^2+(x_1-x_2)^2}$

$$= \sqrt{1+m^2}(x_1-x_2)$$

$$= 2\frac{\sqrt{a^2(1+m^2)-c^2}}{\sqrt{1+m^2}}$$

In a similar manner, we can consider the points of intersection of the straight line $y = mx + c$ with the circle

$$x^2+y^2+2gx+2fy+c=0$$

The straight line $y = mx + a\sqrt{1+m^2}$ is always a tangent to the circle $x^2+y^2=a^2$

The straight line $y = mx + c$ meets the circle in two points which are coincident if $c = a\sqrt{1+m^2}$

But if a straight line meets the circle in two points which are indefinitely close to one another then, it is a tangent to the circle.

The straight line $y = mx + c$ is therefore, a tangent to the circle

If $c = a\sqrt{1+m^2}$

i.e. The equation to any tangent to the circle is

$$y = mx + a\sqrt{1+m^2} \tag{1}$$

Since, the radical on the right hand may have the + or − signs pre-fixed. We see that corresponding to any value of m there are two tangents.

12. **Observation:** The straight line $y = mx + c$ will touch the circle if the $\perp^{lar}$ on it from the origin be equal to a if $\dfrac{c}{\sqrt{1+m^2}} = a$

If $c = a\sqrt{1+m^2}$ .This is however applicable to any other curve be-sides the circle

13. **Normal: The normal at any point P of a curve is the straight line which passes through P and is perpendicular to the tangent at P**

14. **Let us obtain the equation to the normal at the point** (x', y') **of the circle (1)** $x^2 + y^2 = a^2$ **and the circle (2)** $x^2 + y^2 + 2gx + 2fy + c = 0$

   a.   **Case 1:** Circle $x^2 + y^2 = a^2$

      The tangent at (x', y') is

$$xx' + yy' = a^2$$

      i.e. $yy' = a^2 - xx'$

$$y = \frac{-x'}{y'}x + \frac{a^2}{y'} \ (\div \text{ by } y')$$

      The equation to the straight line passing through $(x', y') \perp^{lar}$ to this tangent is

$$y - y' = m(x - x')$$

      Where $m \times \left(\dfrac{-x'}{y'} \right) = -1$

      i.e. $m = \dfrac{y'}{x'}$

The required equation is therefore

$$y - y' = \frac{y'}{x'}(x - x')$$

$$\text{Or} (y - y')x' = y'(x - x')$$

$$\text{Or } yx' - y'x' = y'x - x'y'$$

$$\text{Or } x'y - y'x = 0$$

This straight line passes through the centre of the circle which is the point $(0,0)$

b. **Case 2**: Circle $x^2 + g^2 + 2gx + 2fy + c = 0$

The equation to the tangent at (x', y') to the circle

$$x^2 + g^2 + 2gx + 2fy + c = 0$$

Is $y = \dfrac{-(x' + g)}{y' + f} - \dfrac{(gx' + fy' + c)}{y' + f}$

The equation to the straight line, passing through the points (x', y') and perpendicular to this tangent is

$$y - y' = m(x - m')$$

Where $m \times \left(-\dfrac{x' + g}{y' + f} \right) = -1$

$\text{Or } m = \dfrac{y' + f}{x' + g}$

The equation to the normal is therefore

$$y - y' = \frac{y' + f}{x' + g}(x - x')$$

$$(y - y')(x' + g) = (y' + f)(x - x')$$

$$y(x' + g) - x(y' + f) + fx' - gy' = 0$$

14. **Let us now prove that from any point there can be drawn two tangents, real or imaginary, to a circle.**

Let the equation tothe circle be $x^2 + y^2 = a^2$, and let given point be (x_1, y_1)

The equation to any tangent is $y = mx + a\sqrt{1 + m^2}$

If this pass through the given point (x_1, y_1) we have

$$y_1 = mx_1 + a\sqrt{1 + m^2} \tag{1}$$

This is the equation which gives the values of m corresponding to the tangents which pass through (x_1, y_1)

Now (1) gives

$$y_1 - mx_1 = a\sqrt{1 + m^2}$$

Squaring both sides, we get

$$(y_1 - mx_1)^2 = a^2(1 + m^2)$$

$$y_1^2 + m^2 x_1^2 - 2mx_1 y_1 = a^2 + a^2 m^2$$

Or $m^2(x_1^2 - a^2) - 2mx_1 y_1 + y_1^2 - a^2 = 0 \tag{2}$

The equation (2) is a quadratic equation and gives therefore, two values of m (real, coincident, or imaginary). Corresponding to any given values of x_1 and y_1. For each of these values of m we have corresponding tangent.

The roots of equation (2) are by real, coincident

Or imaginary according as

$(2x_1 y_1)^2 - 4(x_1^2 - a^2)(y_1^2 - a^2)$ is +ve, zero, or −ve according as

$a^2(-a^2 + x_1^2 + y_1^2)$ is +ve, zero, or negative

i.e. according as $x_1^2 + y_1^2 \overset{\geq}{<} a^2$

---

15. **Observation:**If $x_1^2 + y_1^2 > a^2$, the distance of the point (x_1, y_1) from the centre is greater than the radius and hence it lies outside the circle

    If $x_1^2 + y_1^2 = a^2$, the point (x_1, y_1) lies on the circle and the two coincident tangents became the tangent at (x_1, y_1)

    If $x_1^2 + y_1^2 < a^2$, then the point (x_1, y_1) lies within the circle, and no tangents can then be geometrically drawn to the circle. It is however better to say that the tangents are imaginary.

16. **Chord of contact:If from any point T without a circle two tangents TP and TQ be drawn to the circle, the straight line PQ joining the points of contact is called the chord of contact of tangents from T.**

17. **To find the equation of the chord of contact of tangents drawn to the circle** $x^2 + y^2 = a^2$ **from the external point** (x_1, y_1)

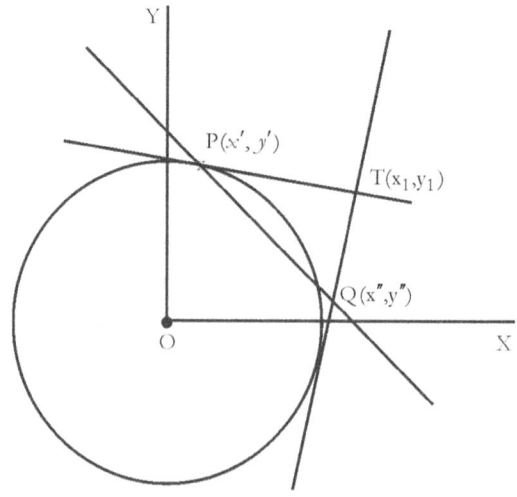

Fig.74

Let T be the point (x_1, y_1) and P and Q the points (x', y') and (x'', y'') respectively.

The tangent at P is

$$xx' + yy' = a^2 \tag{1}$$

And that at Q is

$$xx'' + yy'' = a^2 \tag{2}$$

Since, these tangent pan through T, it co-ordinates (x_1, y_1) must satisfy both (1) and (2)

Hence $x_1 x' + y_1 y' = a^2 \tag{3}$

$$x_1 x'' = y_1 y'' = a^2 \tag{4}$$

The equation to PQ is then

$$xx_1 + yy_1 = a^2 \tag{5}$$

For, since (3) is true, it follows that the point (x'', y'')

i.e. Q lies on (5)

Hence both P and Q lies on the straight line (5) i.e., (5) is the equation to the required chord of contact

18. **Pole and polar definition:** If through a point P there be drawn any straight line to meet the circle in Q and R, the locus of the point of intersection of the tangents at Q and R is called the polar of P, also P is called the pole of the polar.

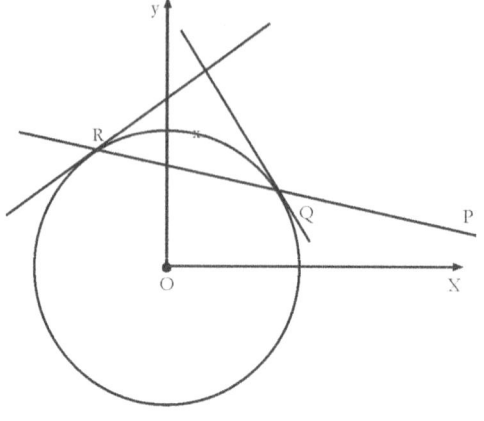

Fig.75

19. Let us obtain the equation to the polar of the point (x_1, y_1)
with respect to the circle $x^2 + y^2 = a^2$

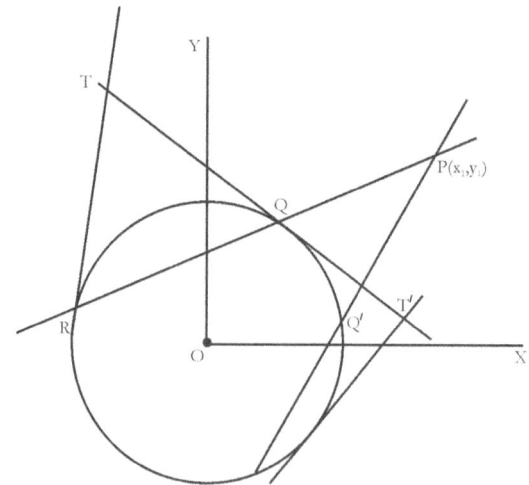

Fig. 76

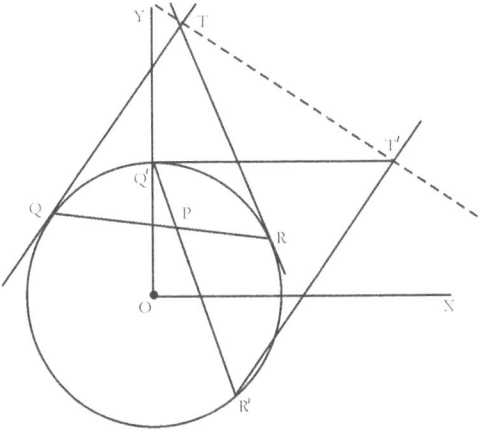

Fig.77

Let QR be any chord drawn through P and let the tangents at Q and R meet in the point T whose co-ordinates are (h, k)

Hence QR is the chord of contact of tangents drawn from the point (h, k) and therefore, its equation is $xh + yk = a^2$

Since, this line passes through the point (x_1, y_1), we have

$$x_1h + y_1k = a^2 \qquad (1)$$

Since, the relation (1) is true it follows that the variable point (h, k) always lies on the straight line whose equation is

$$xx_1 + yy_1 = a^2 \qquad (2)$$

Hence eq (2) is the polar of the point (x_1, y_1)

Note:

In a similar manner it may be proved that the polar of (x_1, y_1) with respect to the circle

$$x^2 + y^2 + 2gx + 2fy + c = 0 \text{ is}$$

$$xx_1 + yy_1 + g(x + x_1) + f(y + y_1) + c = 0$$

20. **Polar: An alternative way to define a polar is as follows. The polar of a given point is the straight line which passes**

through the points of contact of tangents drawn from the given point.

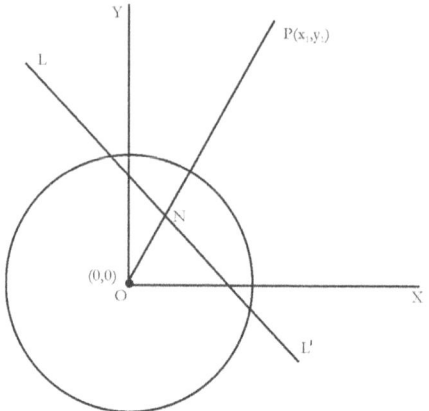

Fig.78

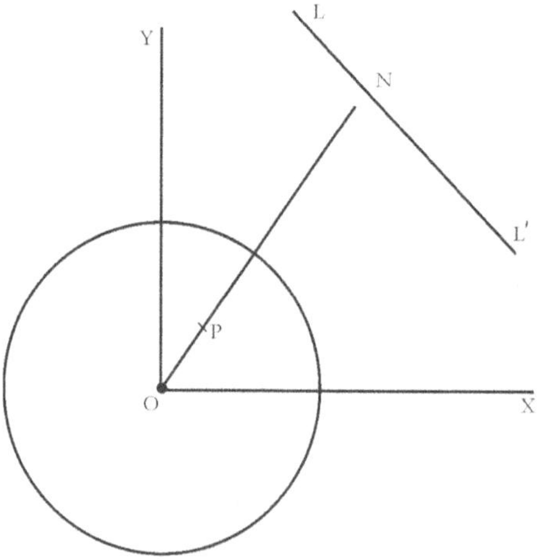

Fig.79

The equation to OP, which is the line joining $(0, 0)$ to (x_1, y_1) is

$$y - 0 = \frac{y_1 - 0}{x_1 - 0}(x - 0)$$

Or $y = \dfrac{y_1}{x_1}x$

Or $xy_1 - x_1y - 0$ (1)

Also the polar of P is $xx_1 + yy_1 = a^2$ (2)

The lines (1) & (2) are $\perp$ to one another

Hence, OP is $\perp$ to polar of P

Also the length of $OP = \sqrt{(x_1 - 0)^2 + (y_1 - 0)^2}$

$\qquad = \sqrt{x_1^{\,2} + y_1^{\,2}}$

And the perpendicular, ON from O upon (2) is

$$ON = \dfrac{a^2}{\sqrt{x_1^{\,2} + y_1^{\,2}}}$$

Hence $OP \cdot ON = \sqrt{x_1^{\,2} + y_1^{\,2}} \times \dfrac{a^2}{\sqrt{x_1^{\,2} + y_1^{\,2}}}$

Or $OP \cdot ON = a^2$

21. **Let us determine the pole of a given line with respect to any circle.**

Let the equation to the given line be

$\qquad Ax + By + C = 0$ (1)

Case 1:

Let the equation to the circle be

$\qquad x^2 + y^2 = a^2$

And let the required pole be (x_1, y_1)

Then, (1) must be the equation to the polar of (x_1, y_1)

i.e. It is the same as the equation

$\qquad xx_1 + yy_1 - a^2 = 0$ (2)

Comparing equation (1) & (2), we have

$$\frac{x_1}{A} = \frac{y_1}{B} = \frac{-a^2}{C}$$

So that $x_1 = \frac{-A}{C}a^2$, and $y_1 = \frac{-B}{C}a^2$

The required pole $= \left(\frac{-A}{C}a^2, \frac{-B}{C}a^2 \right)$

Case 2:

Let the equation to the circle be

$$x^2 + y^2 + 2gx + 2fy + c = 0$$

If (x_1, y_1) be the required pole, then (1) must be equivalent to the equation

$$xx_1 + yy_1 + g(x + x_1) + f(y + y_1) + c = 0$$

Or $x(x_1 + g) + y(y_1 + f) + gx_1 + fy_1 + c = 0$ $\qquad$ (3)

Comparing (1) with (3), we therefore, have

$$\frac{x_1 + g}{A} = \frac{y_1 + f}{B} = \frac{gx_1 + fy_1 + c}{c}$$

By solving these equations we have the values of x_1 and y_1

i.e. $\dfrac{x_1}{A} - \dfrac{gx_1}{C} = \dfrac{fy_1 + C}{C} - \dfrac{g}{A}$

$$x_1 \left(\frac{C - g}{AC} \right) = \frac{A(fy_1 + C) - gC}{CA}$$

$$x_1 = \frac{Afy_1 + AC - gC}{C - g}$$

Also $\dfrac{y_1}{B} - \dfrac{fy_1}{C} = \dfrac{gx_1 + C}{C} - \dfrac{f}{B}$

$$y_1 \left(\frac{C - (-f)}{BC} \right) = \frac{Bgx_1 + BC - CF}{BC}$$

$$y_1 = \frac{Bgx_1 + BC - Cf}{C - f}$$

22. **To find the length of the tangent that can be drawn from the point** (x_1, y_1) **tothe circles (1)** $x^2 + y^2 = a^2$ **(2)** $x^2 + y^2 + 2gx$ $+ 2fy + c = 0$

If T be an external point, TQ a tangent and O be the center of the circle, then TQO is a right angle and hence

$$TQ^2 = OT^2 - OQ^2$$

Case 1:

If the equation to the circle to be

$x^2 + y^2 = a^2$, O is the origin

$OT^2 = x_1^2 + y_1^2$ and $OQ^2 = a^2$

Hence $TQ^2 = x_1^2 + y_1^2 - a^2$

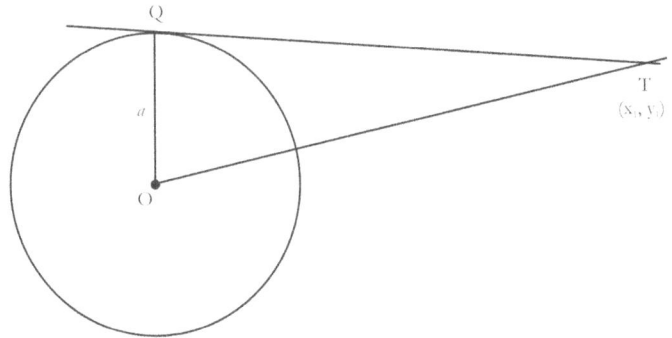

Fig.80

Case 2:

Let the equation to the circle be

$$x^2 + y^2 + 2gx + 2fy + c = 0$$

i.e. $(x + g)^2 + (y + f)^2 = g^2 + f^2 - c$

In this case O is the point $(-g, -f)$ and

$$OQ^2 = g^2 + f^2 - c$$

$$\text{Hence } OT^2 = [x_1 - (-g)]^2 + [y_1 - (-f)]^2$$

$$= (x_1 + g)^2 + (y_1 + f)^2$$

Therefore

$$TQ^2 = (x_1 + g)^2 + (y_1 + f)^2 - (g^2 + f^2 - c)$$

$$= x_1^2 + y_1^2 + 2gx_1 + 2fy_1 + c$$

23. Let us obtain the equation to the pair of tangents that can be drawn from the point (x_1, y_1) to the circle $x^2 + y^2 = a^2$

Let (h, k) be any point on either of the tangents from (x_1, y_1). Since, any straight line touches a circle if the perpendicular on it from the centre is equal to the radius the perpendicular from the origin upon the line joining (x_1, y_1) to (h, k) must be equal to a.

The equation to the straight line joining these two point is

$$y - y_1 = \frac{k - y_1}{h - x_1}(x - x_1)$$

Or $(y - y_1)(h - x_1) = (k - y_1)(x - x_1)$

i.e. $y(h - x_1) - x(k - y_1) + kx_1 - hy_1 = 0$

Hence, $\dfrac{kx_1 - hy_1}{\sqrt{(h - x_1)^2 + (k - y_1)^2}} = a$

So that, $(kx_1 - hy_1)^2 = a^2[(h - x_1)^2 + (k - y_1)^2]$

Therefore, the point (h, k) always lies on the locus

$$(x_1y - xy_1)^2 = a^2[(x - x_1)^2 + (y - y_1)^2] \tag{1}$$

This is therefore, is the required condition

Above eq, may be written in the form

$$x^2(y_1^2 - a^2) + y^2(x_1^2 - a^2) - a^2(x_1^2 + y_1^2)$$

$$= 2xyx_1y_1 - 2a^2xx_1 - 2a^2yy_1$$

i.e. $(x^2 + y^2 - a^2)(x_1^2 + y_1^2 - a^2) = x^2 x_1^2 + y^2 y_1^2 + a^4 + 2xy x_1 y$

$$-2a^2 xx_1 - 2a^2 yy_1 = (xx_1 + yy_1 - a^2)^2 \qquad (2)$$

Similarly, the equation to the pair of tangents that can be drawn from (x_1, y_1) to the circle

$$(x - f)^2 + (y - g)^2 = a^2$$

$$\{(x - f)^2 + (y - g)^2 - a^2\}\{(x_1 - f)^2 + (y_1 - g)^2 - a^2\}$$

$$= \{(x - f)(x_1 - f) + (y - g)(y_1 - g) - a^2\}^2 \qquad (1)$$

If the equation to the tangents is similarly

$$(x^2 + y^2 + 2gx + 2fy + c)(x_1^2 + y_1^2 + 2gx_1 + 2fy_1) + c$$

$$= [(xx_1 + yy_1 + g(x + x_1) + f(y + y) + c]^2 \qquad (2)$$

24. Let us determine the general equation of a circle referred to polar co-ordinates.

Let O be the origin or Pole OX the initial line, C the centre and a the radius of the circle.

Le the polar co-ordinates of C be R and α, so that $OC = R$ and $\underline{XOC} = \alpha$

Let a radius vector through O at on angle θ with the initial line cut the circle in P and Q. Let OP or OQ be r

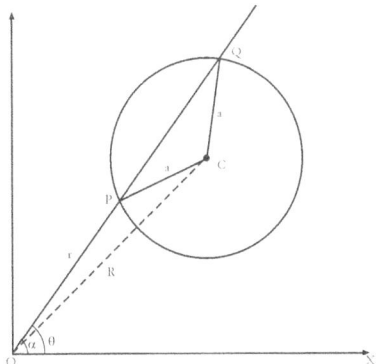

Fig. 81

Then, we have

$$CP^2 = OC^2 + OP^2 - 2OC \cdot OP \cos COP$$

i.e. $a^2 = R^2 + r^2 - 2Rr \cos(\theta - \alpha)$

i.e. $r^2 - 2Rr \cos(\theta - \alpha) + R^2 - a^2 = 0$ (1)

Is the required polar equation

1:

Let the initial line be taken to go through the centre C. Then $\alpha = 0$

Equation is $r^2 - 2Rr \cos\theta + R^2 - a^2 = 0$

Case 2:

Let the pole O be taken on the circle, so

$$R = O \ \ C = a$$

Equation is $r^2 - 2ar \cos(\theta - \alpha) = 0$ ($\div$ by r)

Or $r = 2a \cos(\theta - \alpha)$

Case 3:

Let the pole be on the circle and also let the initial line pass through the centre of the circle

Then $\alpha = 0, R = a$

The general equation reduces then to the simple form $r = 2a \cos\theta$

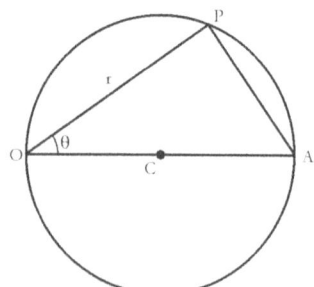

Fig.82

For if OCA be a diameter,

We have $OP = OA \cos\theta$

$$r = 2a \cos\theta$$

24. Let us now determine the general equation to a circle when the reference axes are oblique, which meet at an angle w.

Let C be the centre and a the radius of the circle. Le the co-ordinate of C be (h, k) so that if CM, drawn parallel to the axis of y, meets OX in M, the

$$OM = h \text{ and } MC = k$$

Let P be any point on the circle whose co-ordinates are x and y. Draw PN, the ordinate of P. and CL parallel to OX to meet PN in L.

Then, $CL = MN = ON - OM = x - h$

And $LP = NP - LN = NP - CM = y - k$

Also $\lfloor CLP = \lfloor ONP = 180° - \lfloor PNX$

$$= 180° - w$$

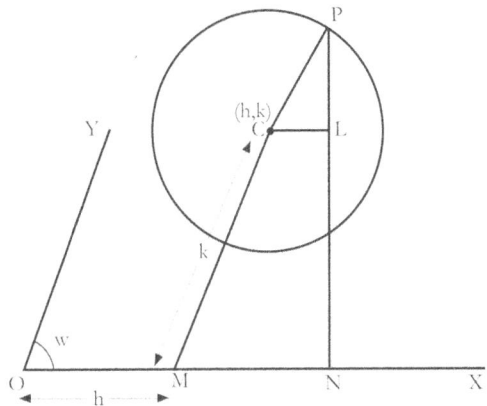

Fig.83

Hence, since $CL^2 + LP^2 - 2CL \cdot PL \cdot \cos CLP = a^2 = CP^2$

i.e. $(x - h)^2 + (y - k)^2 + 2(x - h)(y - k)\cos w = a^2$

Or $x^2 + y^2 - 2xh + y^2 + k^2 - 2yk + 2xy \cos w$

$$-2xk\cos w - 2hy\cos w + 2hk\cos w = a^2$$

$$\text{Or } x^2 + y^2 + 2xy\cos w - 2x(h + k\cos w)$$

$$-2y(k + h\cos w) + h^2 + k^2 + 2hk\cos w = a^2$$

25. To find the equation to the straight line joining two points α **and** β **, on the circle** $x^2 + y^2 = a^2$

Let P and Q be the points and let ON be the $\perp^{lar}$ drawn from the origin, on the straight line PQ; then ON bisects the angle POQ, and hence

$$\lfloor XON = \frac{1}{2}\left(\lfloor XOP + \lfloor XOQ\right) = \frac{1}{2}(\alpha + \beta)$$

Also $ON = OP\cos NOP = a\cos\left(\dfrac{\alpha - \beta}{2}\right)$

The equation to PQ is therefore

$$x\cos\left(\frac{\alpha + \beta}{2}\right) + y\sin\left(\frac{\alpha + \beta}{2}\right) = a\cos\left(\frac{\alpha - \beta}{2}\right)$$

If we put $\beta = \alpha$, we have the equation to the tangent

$$x\cos\alpha + y\sin\alpha = a$$

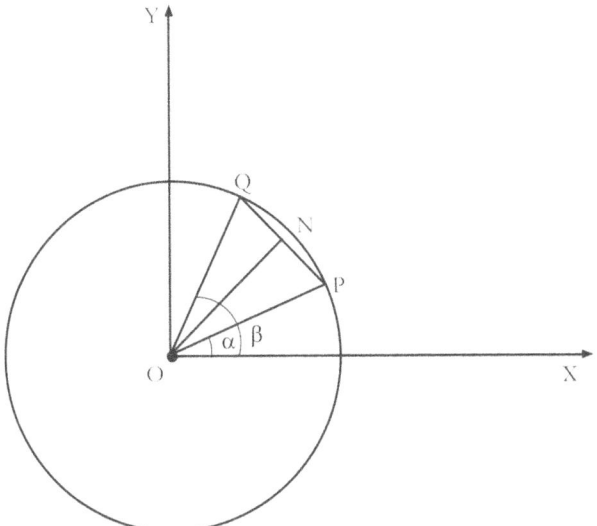

Fig. 84

If the equation to the circle be in more general form

$$(x-h)^2 + (y-k)^2 = a^2$$

We may express the co-ordinates of P in the form

$$(h + a\cos\alpha, k + a\sin\alpha)$$

These values satisfies above equations

The equation to the straight line joining the points α and β is

$$(x-h)\cos\left(\frac{\alpha+\beta}{2}\right) + (y-k)\sin\left(\frac{\alpha+\beta}{2}\right) = a\cos\left(\frac{\alpha+\beta}{2}\right)$$

And so the tangent at the point α is

$$(x-h)\cos\alpha + (y-k)\sin\alpha = a$$

26. **Common tangents to two circles**: If O_1 and O_2 be the centre of two circles whose radii are r_1 and r_2 and if one pair of common tangents meet O_1O_2 in T_1 and the other pair meet it in T_2, then, by similar triangles, we have

$$\frac{O_1T_2}{T_2O_2} = \frac{r_1}{r_2} = \frac{O_1T_1}{O_2T_1}$$

The points T_1 and T_2 therefore, divide $O_1 O_2$ in the ratio of the radii.

Solved Examples

Example 1 : Find the equation to the circle, whose center, is the point $(-3,4)$ and whose radius is 7, is

$$(x-h)^2 + (y-k)^2 = a^2$$

$$[x-(-3)]^2 + (y-4)^2 = 7^2$$

$$(x+3)^2 + (y-4)^2 = 49$$

$$x^2 + 6x + 9 + y^2 - 8y + 16 - 49 = 0$$

$$x^2 + y^2 + 6x - 8y - 24 = 0$$

Example 2 : Find center and radius of the circle
$$x^2 + y^2 + 4x - 6y = 0$$

Compare $x^2 + y^2 + 4x - 6y = 0$ with

$$x^2 + y^2 + 2gx + 2fy + c = 0$$

We have $2g = 4, 2f = -6, c = 0$ $g = 2$ $f = -3$

Centre is $(-g, -f) = (-2, 3)$

$$\text{Radius} = \sqrt{g^2 + f^2 - c} = \sqrt{4+9}$$

$$= \sqrt{13}$$

Example 3: Show that $45x^2 + 45y^2 - 60x + 36y + 19 = 0$ represents equation of circle, hence find its center and radius.

Dividing throughout the equation by 45, we get

$$x^2 + y^2 - \frac{4}{3}x + \frac{4}{5}y + \frac{+19}{45} = 0$$

Compare with $x^2 + y^2 + 2gx + 2fy + c =$

We have $2g = \dfrac{-4}{3}, 2f = \dfrac{4}{5}\ c = \dfrac{19}{45}$

$$g = \dfrac{-4}{6} = \dfrac{-2}{3}\ f = \dfrac{2}{5}$$

Centre $(-g, -f) = \left(\dfrac{2}{3}, \dfrac{-2}{5} \right)$

$$r = \sqrt{g^2 + f^2 - c} = \sqrt{\dfrac{4}{9} + \dfrac{4}{25} - \dfrac{19}{45}}$$

$$= \sqrt{\dfrac{41}{225}} = \dfrac{\sqrt{41}}{15}$$

Example 4: Find the equation to the circle which passes through the points $(1,0)(0,-6)$ and $(3,4)$.

Let the equation to the circle be

$$x^2 + y^2 + 2gx + c = 0 \tag{1}$$

Since, $(1, 0)$ passing through (1), we have

$$1^2 + 0 + 2g(1) + 2f(0) + c = 0$$

i.e. $2g + c = -1$ \hfill (2)

$(0, -6)$ passing through (1) gives

$$0 + 36 + 2g(0) + 2f(-6) + c = 0$$

Or $-12f + c = -36$ \hfill (3)

Again $(3, 4)$ passing through (1) gives

$$9 + 16 + 2g(3) + 2f(4) + c = 0$$

Or $6g + 8f + c = -25$ \hfill (4)

(3) − (2) gives

$$-12f + c = -36$$
$$2g + c = -1$$
$$\underline{(-)\,(-)\,\ (+)}$$
$$-2f - 2g = -36$$
$$12f + 2g = 36 \tag{5}$$

(4) - (3) gives

$$6g + 8g + c = -25$$
$$-2f + c = -36$$
$$\underline{(+)\qquad (-)\ \ (+)}$$
$$6g + 20f = 11 \tag{6}$$

$3 \times (5) - (6)$ gives

$$36f + 6g = 35 \times 6$$
$$20f + 6g = 11$$
$$\underline{(-)\ (-)\ (-)}\qquad \text{and } g = \dfrac{-71}{4}$$
$$16f = 210 - 11$$

From (2) $c = \dfrac{69}{2}$

Or $f = \dfrac{47}{8}$

Substituting these values in (1), the required equation is

$$4x^2 + 4y^2 - 142x + 47y + 138 = 0$$

Example 5: Find the equation to the circle which touches the axis of y at a distance +4 from the origin and cuts off an intercept 6 from the axis of x.

Any circle can be represented by the equation $x^2 + y^2 + 2gx + 2fy + c = 0$

This meets the axis of y in points given by

$$y^2 + 2fy + c = 0 \tag{1}$$

The roots of this equation must be equal and each equal to 4, so that it must be equivalent to $(y-4)^2 = 0$

Hence $y^2 + 16 - 8y = 0$ (2)

Comparing (1) & (2)

$$2f = -8 \text{ or } f = -4 \text{ and } c = 16$$

The equation to the circle is then

$$x^2 + y^2 + 2gx - 8y + 16 = 0$$

This meets the axis of x in points given by

$$x^2 + 2gx + 16 = 0$$
$$(\because c = 16$$
$$y = f = 0$$

i.e. at points distant

$$-g + \sqrt{g^2 - 16} \text{ and } -g - \sqrt{g^2 - 16}$$

Hence $6 = 2\sqrt{g^2 - 16}$

Therefore $3 = \sqrt{g^2 - 16}$

Or $g^2 - 16 = 9$

$$g^2 = 25 \Rightarrow g = \pm 5$$

And the required equation is

$$x^2 + y^2 \pm 10x - 8y + 16 = 0$$

Example 6: Find the equations to the tangents to the circle $x^2 + y^2 - 6x + 4y = 12$ which are parallel to the straight line $4x + 3y + 5 = 0$.

Any straight line parallel to the given eq $4x + 3y + 5 = 0$

Is $4x + 3y + c = 0$ (1)

The equation to the circle is

$$x^2 - 6x + 9 + y^2 + 4y + 4 = 12 + 9 + 4$$

Or $(x-3)^2 + (y+2)^2 = 5^2$

The straight line (1), if it be a tangent, must be therefore such that its distance from the point $(3,-2)$ is equal to $\cdots 5$.

Hence $\dfrac{12 - 6 + c}{\sqrt{4^2 + 3^2}} = \pm 5$

$$\therefore \frac{ax_1 + 6y_1 + c}{\sqrt{a^2 + b^2}} = \pm r$$

So that

$$c = -6 \pm 25$$
$$= 19 \text{ or } -31$$

The required tangents are therefore

$$4x + 3y + 19 = 0 \text{ and } 4x + 3y - 31 = 0$$

Example 7: Find the pole of the straight line $9x + y - 28 = 0$, with respect to the circle $2x^2 + 2y^2 - 3x + 5y - 7 = 0$

$9x + y - 28 = 0$ \hspace{2cm} (1)

$2x^2 + 2y^2 - 3x + 5y - 7 = 0$ \hspace{1.5cm} (2)

If (x_1, y_1) be the required point the line (1) must coincide with the polar of (x_1, y_1), whose equation is

$$2xx_1 + 2yy_1 - \frac{3}{2}(x + x_1) + \frac{5}{2}(y + y_1) - 7 = 0$$

i.e. $x\left(2x_1 - \dfrac{3}{2}\right) + y\left(2y_1 + \dfrac{5}{2}\right) - \dfrac{3}{2}x_1 + \dfrac{5}{2}y_1 - 7 = 0$

Or $x(4x_1 - 3) + y(4y_1 + 5) - 3x_1 + 5y_1 - 14 = 0$
\hspace{2cm} (3)

Since (1) and (3), are the same

We have

$$\frac{4x_1 - 3}{9} = \frac{4y_1 + 5}{1} = \frac{-3x_1 + 5y_1 - 14}{-28}$$

Hence $\dfrac{4x_1 - 3}{9} = \dfrac{-3x_1 + 5y_1 - 14}{-28}$

Or $-28(4x_1 - 3) = 9(-3x_1 + 5y_1 - 14)$

$$-112x_1 + 84 = -27x_1 + 45y_1 - 126$$

Or $-85x_1 = 45y_1 - 210$

Or $-17x_1 = 9y_1 - 42$

And $-28(4y_1 + 5) = -3x_1 + 5y_1 - 4$

$$-112y_1 - 140 = -3x_1 + 5y_1 - 14$$

Or $3x_1 = 117y_1 + +126 = 0$

We have $17x_1 + 9y_1 = 42 \times 13$

$$3x_1 - 117y_1 = 126$$

$$221x + 117y_1 = 546$$

$$\underline{3x_1 - 117y_1 = 126}$$

$$224x_1 = 672$$

$$\therefore x_1 = \frac{672}{224}$$

$$x_1 = 3$$

Again $3x_1 - 117y_1 = 126$

$$117y_1 = 3x_1 - 126$$

$$= 3(3) - 126$$

$$117y_1 = -117$$

Or $y_1 = -1$

Required point $= (3, -1)$

Example 8: If the axes be inclined at $60°$, prove that the equation $x^2 + xy + y^2 - 4x - 5y - 2 = 0$ represents a circle and find its center and radius.

$$x^2 + xy + y^2 - 4x - 5y - 2 = 0 \quad (1)$$

If w be equal to $60°$, so that $\cos w = \cos 60 = \dfrac{1}{2}$

Equation is $x^2 + y^2 + 2xy \cos w - 2x(h + k \cos w)$

$$-2y(k + h \cos w) + h^2 + k^2 + 2hk \cos w = a^2$$

Or $x^2 + y^2 + xy - x(2h + k) - y(2k + h) + h^2 + k^2 + hk = a^2$

This equation agrees with (1) if

$$2h + k = 4 \qquad\qquad (2)$$
$$2k + h = 5 \qquad\qquad (3)$$
$$2 \times (2) - (3)$$
$$4h + 2k = 8$$
$$h + 2k = 5$$
$$\underline{(-)(-)\ (-)}$$
$$3h = 3$$

Or $h = 1$

From (2) $k = 4 - 2h$

$$= 4 - 2$$
$$k = 2$$

Equation (4) then gives

$$a^2 = h^2 + k^2 + hk + 2$$
$$= 1^2 + 2^2 + 1(2) + 2$$
$$= 1 + 4 + 4$$
$$= 9$$
$$\therefore a = 3$$

The equation (1) therefore, represents a circle whose centre is the point (1, 2) and whose radius is 3, the axes being inclined at $60°$

Co-ordinates of a point on a circle expressed in terms of one single variable:

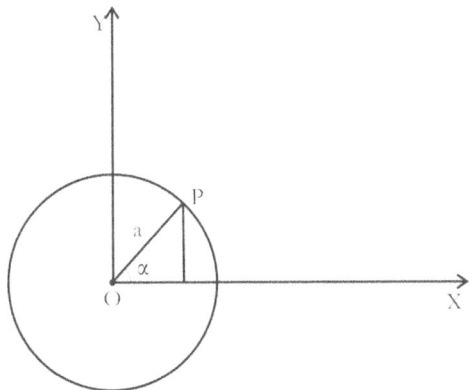

Fig.85

The co-ordinates of the point P are

$$x = a\cos\alpha \text{ and } y = a\sin\alpha$$

Which satisfies the equation

$$x^2 + y^2 = a^2$$

$$(a\cos\alpha)^2 + (a\sin\alpha)^2 = a^2(\cos^2\alpha + \sin^2\alpha)$$

$$= a^2$$

Example 9: Find the four common tangents to the circles

$$x^2 + y^2 - 22x + 4y + 100 = 0 \text{ and } x^2 + y^2 + 22x - 4y - 100 = 0$$

Comparing (1) and (2) with $x^2 + y^2 + 2gx + 2fy + c = 0$

Centre of (1) is $(11, -2)$ & $r = \sqrt{121 + 4 - 100} = \sqrt{25} = 5$

Center of (2) is $(-11, 2)$ & $r = \sqrt{121 + 4 + 100} = \sqrt{225} = 15$

Then T_2 is the point dividing internally the line joining the centre in the ratio 5:15 and its co-ordinates are

$$A(11,-2) \quad B(-11,2) \quad 5:15$$
$$\underset{x_1 \ y_1}{} \qquad \underset{x_2 \ y_2 \ m:n}{}$$

$$T_2(x, y) = \left(\frac{mx_2 + nx_1}{m+n}, \frac{my_2 + ny_1}{m+n} \right)$$

$$= \left[\frac{5(-11)+15(11)}{5+15}, \frac{5(2)+15(-2)}{5+15} \right]$$

$$= \left(\frac{11}{2}, -1 \right)$$

Similarly T_1 is the point dividing this line externally in the ratio 5:15, and hence its co-ordinates are

$$T_1(x, y) = \left(\frac{15 \times 11 - (5 \times -11)}{15 - 5}, \frac{15 \times (-2) - 5 \times 2}{15 - 5} \right)$$

$$= (22, -4)$$

Let the equation to either of the tangents passing through T_2 be

$$y + 1 = m\left(x - \frac{11}{2} \right)$$

Then the perpendicular from the point $(11, -2)$ on it is equal to ± 5, and hence

$$\frac{m\left(11 - \frac{11}{2} \right) - (-2+1)}{\sqrt{1 + m^2}} = \pm 5$$

$$\frac{m\left(\frac{11}{2} \right) + 1}{\sqrt{1 + m^2}} = \pm 5$$

$$\frac{11m + 2}{2} = \pm 5\sqrt{1 + m^2}$$

$$11m + 2 = \pm 10\sqrt{1 + m^2}$$

On solving we get

$$(11m + 2)^2 = 10(m^2 + 1)$$

$$121m^2 + 4 + 44m = 10m^2 + 10$$

$$111m^2 + 44m - 6 = 0$$

We get $m = -\dfrac{24}{7}$ or $\dfrac{4}{3}$

The required tangents through T_2 are therefore

$$24x + 7y = 125 \ 24\text{and} \ 4x - 3y = 25$$

Similarly, the equations to the tangents through T_1 is

$$y + 4 = m(x - 22)$$

Where $\dfrac{m(11 - 22) - (-2 + 4)}{\sqrt{1 + m^2}} = \pm 5$

On solving, we have $m = \dfrac{7}{24}$ or $\dfrac{-3}{4}$

On substitution in (2), the required equations are therefore,

$$7x - 24y = 250 \text{ and } 3x + 4y = 50$$

Example 10: Find the locus of a point P which moves so that its distance from a given point O is always in a given ratio n:1 to its distance from another given point A.

Take O as origin and the direction of OA as the axis of x

Let the distance OA be a, so that A is the point $(a, 0)$

If (x, y) be the co-ordinates of any position of P, we have

$$OP^2 = n^2 \cdot AP^2$$

i.e. $x^2 + y^2 = n^2[(x - a)^2 + y^2]$

i.e. $(x^2 + y^2)(n^2 - 1) - 2an^2 x + n^2 a^2 = 0$ \hfill (1)

Hence the locus of P is a circle

Let this circle meet the axis of x in the points C and D.

Then OC and OD are the roots of the equation obtained by putting y equal to zero in (1)

Hence $OC = \dfrac{na}{n+1}$ and $OD = \dfrac{na}{n-1}$

We therefore, have

$$CA = \dfrac{a}{n+1} \text{ and } AD = \dfrac{a}{n-1}$$

Hence $\dfrac{OC}{CA} = \dfrac{OD}{AD} = n$

The points C and D therefore, divide the line OA in the given ratio, and the required circle is on CD as diameter.

9. The Parabola

1. **Conic Section:**The locus of a point P, which moves so that its distance from a fixed point is always in a constant ratio to its perpendicular distance from a fixed straight line, is called a Conic Section.

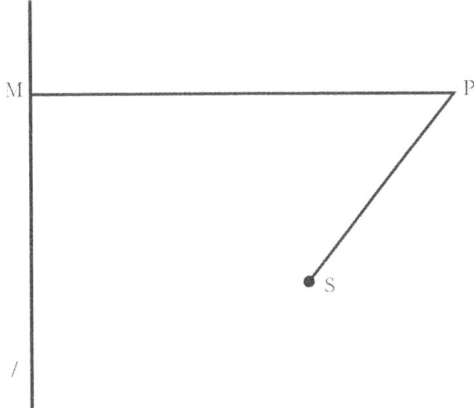

Fig. 86

    a. The **fixed point** is called the **Focus** and is usually denoted by S.

    b. The **constant ratio** is called the **Eccentricity** and is denoted by e.

    c. The **fixed straight line** is called the **Directrix.**

    d. The **straight line passing through the Focus** and **perpendicular to the Directrix** is called the **Axis.**

    e. When the Eccentricity e is **equal to unity** the conic section is called a **parabola**

    f. When the Eccentricitye is **less than unity**, it is called on **Ellipse**

    g. When the Eccentricitye is **greater than unity**, it is called a **Hyperbola.**

2. We will focus our attention on parabola. Let us recall that the eccentricity of this conic section is equal to 1.

3. **Let us now determine the equation of a parabola.**

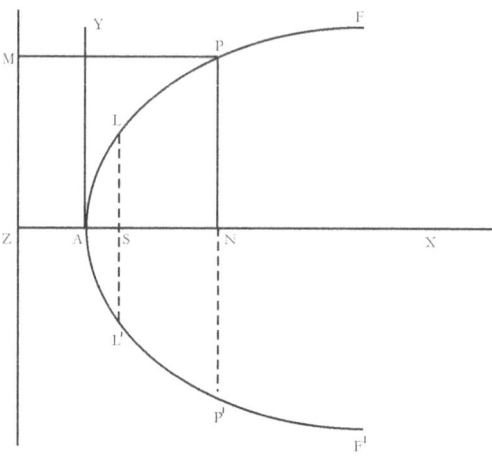

Fig. 87

Let S be the fixed point and ZM the directrix. We require, the locus of a point P which moves so that its distances from S is always equal to PM its perpendicular distance form ZM

$$SP = PM$$

Draw SZ⊥ to the directrix and bisect SZ in the point A, produce ZA to X.

The point A is clearly a point on the curve and is called the vertex of the parabola

Take A as origin and AX-axis of X and AY⊥ to it

Let $ZA = AS = a$,

Let P be (x, y)

Join SP and draw PN & PM⊥ to the axis and directrix

We have then $SP^2 = PM^2$

$$(x - a)^2 + y^2 = (a + x)^2$$

$$S(a,0)P(x,y)$$
$$SP = \sqrt{(x-a)^2 + (y-0)^2}$$
$$PM = ZN$$
$$= ZA + AN$$
$$= a + x$$
$$\because AN = x \ \& \ PN = y$$
$$x^2 + a^2 - 2xa + y^2 = a^2 + x^2 + 2ax$$

Or $y^2 = 4ax$

Is the required eq of parabola

3. Observations:

Case 1:

If instead of AX and AY we take the axis and the directrix as the axis of co-ordinates, the equation would be $(x - 2a)^2 + y^2 = x^2$

Or $x^2 + 4a^2 - 4ax + y^2 = x^2$

$$y^2 = 4a(x - a)$$

Case 2:

Similarly, if the axis SX and a $\perp$ lin SL be taken as the axis of co-ordinates, the equation is

$$x^2 + y^2 = (x + 2a)^2$$

Or $x^2 + y^2 = x^2 + 4a^2 + 4ax$

Or $y^2 = 4a(a + x)$

Summary:

These two equations may be deduced from the equation of the previous article by transforming the origin, firstly to the point $(-a,0)$ and secondly to the point $(a,0)$

4. Let us understand the conditions for determining where an arbitrary point is located with respect to a parabola. It can be

shown that the quantity $y'^2 - 4ax'$ is negative, zero, or positive according as the point (x', y') is within, upon or without the parabola.

Let Q be the point (x', y') and let it be within the curve. i.e. be between the curve and the axis AX. Draw the ordinate QN and let it meet the curve in P.

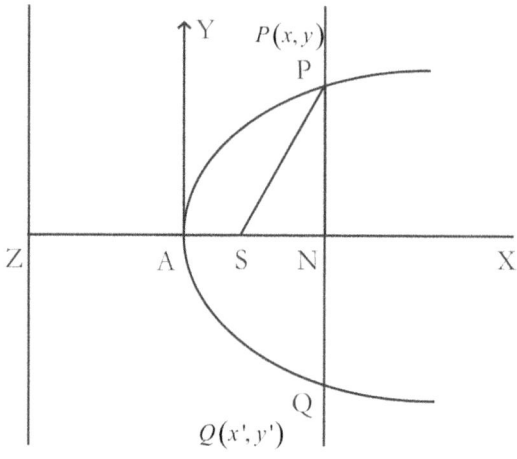

Fig. 88

Then $PN^2 = 4a \cdot x'$

Hence $y'^2 < PN^2 \Rightarrow y'^2 < 4ax'$

i.e. $QN^2 < PN^2$

And hence is $< 4ax'$

$\therefore y'^2 - 4ax'$ is negative

Similarly, if Q be without the curve, then $y'^2 > 4ax'$

i.e. QN^2 is $> PN^2$, and hence is $y'^2 > 4ax'$

5. Latus Rectum: The latus rectum of any conic is the double ordinatedrawn through the focus S.

In the case of the parabola we have SL = distance of L from the directrix $= SZ = 2a$

Hence, the lotus rectum $= 4a$

The quantity $4a$ is also after called the principal parameter of the curve.

6. **Focal distance**: Focal distance of any point P is the distance SP. This focal distance $= PM = ZN = ZA + AN = a + x$

7. **Let us determine the points of intersection of any straight line with the parabola $y^2 = 4ax$**

Let the parabola be $y^2 = 4ax$ (1)

The equation to any straight line is

$$y = mx + c \qquad (2)$$

Substituting the value of y from (2) in (1), we have

$$(mx + c)^2 = 4ax$$

$$m^2x^2 + c^2 + 2mcx = 4ax$$

Or $m^2x^2 + 2x(mc - 2a) + c^2 = 0$ (3)

This is a quadratic equation for x and therefore, has two roots, real, coincident, or imaginary.

The roots of (3) are real or imaginary according as $\{2(mc - 2a)\}^2$

$-4m^2c^2$ is positive or negative

i.e. $4m^2c^2 + 16a^2 - 16mca - 4m^2c^2$ is +ve or −ve

Or $a^2 - mca$ is +ve or −ve

Or $a - mc$ is +ve or −ve

Or $mc \geq a$

8. **Let us determine the length of the chord intercepted by the parabola on the straight line $y = mx + c$**

If (x_1, y_1) and (x_2, y_2) be the common points of intersection then

---

We have $m^2x^2 + 2x(mc - 2a) + c^2 = 0$

$$(x_1 - x_2)^2 = (x_1 + x_2)^2 - 4x_1x_2$$

$$\because x_1 + x_2 = \frac{-2(mc - 2a)}{m^2}$$

$$x_1x_2 = \frac{c^2}{m^2}$$

$$= \left[\frac{-2(mc - 2a)}{m^2}\right]^2 - 4\left[\frac{c^2}{m^2}\right]$$

$$= \frac{4(mc - 2a)^2}{m^4} - \frac{4c^2}{m^2}$$

$$= \frac{4(m^2c^2 + 4a^2 - 4amc) - 4m^2c^2}{m^4}$$

$$= \frac{16a^2 - 16amc}{m^4}$$

$$= \frac{16a(a - mc)}{m^4}$$

And $y_1 - y_2 = m(x_1 - x_2)$

Hence the required length $= \sqrt{(y_1 - y_2)^2 + (x_1 - x_2)^2}$

$$= \sqrt{[m(x_1 - x_2)]^2 + (x_1 - x_2)^2}$$

$$y_1 - y_2 = (x_1 - x_2)\sqrt{m^2 + 1}$$

$$= \frac{4\sqrt{a(a - mc)}}{m^2}\sqrt{1 + m^2}$$

9. **To find the equation to the tangent at any point (x', y') of the parabola $y^2 = 4ax$**

Let P be the point (x', y') and Q a point (x'', y'') on the parabola
The equation to the line PQ is

$$y - y' = \frac{y'' - y'}{x'' - x'}(x - x') \tag{1}$$

Since, P and Q both lie on the curve, we have

$$y'^2 = 4ax' \tag{2}$$

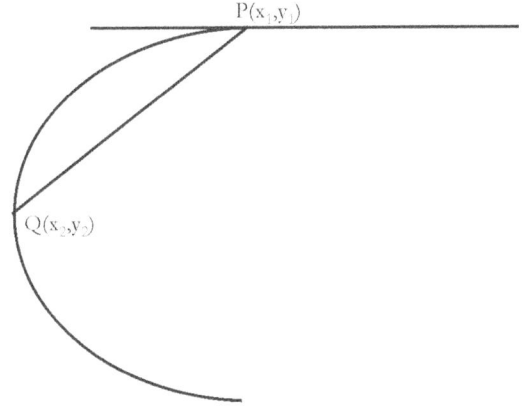

Fig. 89

And $y''^2 = 4ax'' \tag{3}$

Hence, by subtraction, we have

$$y''^2 - y'^2 = 4ax'' - 4ax'$$
$$= 4a(x'' - x')$$
$$(y'' + y')(y'' - y') = 4a(x'' - x')$$

And hence $\dfrac{y'' - y'}{x'' - x'} = \dfrac{4a}{y'' + y'}$

Substituting this value in equation (1), we have, as the equation at PQ

$$y - y' = \frac{4a}{y'' + y'}(x - x')$$
$$(y - y')(y'' + y') = 4a(x - x')$$
$$y(y'' + y') = 4ax + y'y'' + y'^2 - 4ax'$$

$$(\because y'^2 - 4ax' = 0)$$
$$= 4ax + y'y'' \tag{4}$$

To obtain the equation of the tangent at (x', y') we take Q indefinitely close to P, and hence in the limit put $y'' - y'$

The equation (4) then becomes

$$y(y' + y') = 4ax + y'^2$$

Or $2\,yy' = 4ax + 4ax'$

$$2\,yy' = 4a(x + x')$$

Or $yy' = 2a(x + x')$

10. Let us obtain an equation to the normal to a parabola at point $P(x', y')$

The required normal is the straight line which passes through the point (x', y') and is $\perp$ to the tangent

i.e. To the straight line

$$y = \frac{2a}{y'}(x + x')$$

Its equation is therefore

$$y - y' = m'(x - x')$$

Where $m' \times \dfrac{2a}{y'} = -1$

i.e. $m' = \dfrac{-y'}{2a}$

And the equation to the normal is

$$y - y' = \frac{-y'}{2a}(x - x') \tag{1}$$

11. Let us express the equation of the normal in the form-

$$y = mx - 2am - am^2$$

In equation (1) of item 10.,weput

$$\frac{y'}{2a} = m$$

i.e. $y' = -2am$

Hence $x' = \dfrac{y'^2}{4a} = am^2$

The normal is therefore

$$y + 2am = m(x - am^2)$$

i.e. $y = mx - 2am - am^3$

And it is normal at the point $(am^2, -2am)$ of the curve.

11. **Sub-tangent and subnormal:** If the tangent and normal at any point P of a conic section meet the axis in T & G respectively and PN be the ordinate at P, then NT is called the sub-tangent and NG the subnormal of P

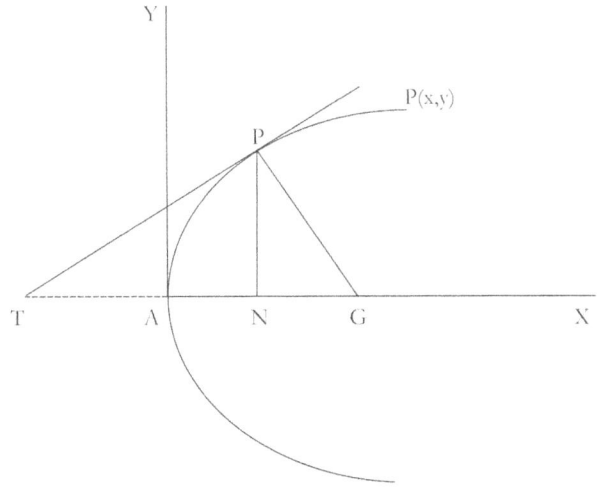

Fig. 90

12. Let us determine the lengths of the sub-tangent and subnormal.

If P be the point (x', y'), the equation to TP is

$$yy' = 2a(x + x') \qquad (1)$$

To obtain the length of AT, we have to find the point where this straight line meets the axis x. i.e. we put $y = 0$ in (1)

And we have $x = -x'$ $\qquad (2)$

Hence $AT = AN$

Hence, the sub-tangent $NT = 2AN =$ twice the abscissa of the point P.

Since TPG, is a right-angled Δ^{le}, we have

$$PN^2 = TN \cdot NG$$

Hence, the subnormal $NG = \dfrac{PN^2}{TN} = \dfrac{PN^2}{2AN} = 2a$

13. Let us now turn our attention to identifying a few properties of a parabola.

1. If the tangent and normal at any point P of the parabola meet the axis in T and G respectively, them $ST = SG = SP$, and the tangent at P is equally inclined to the axis and the focal distance of P.

    Let P be the point (x', y')

    Draw PM $\perp$ to the directix.

    We have $TA = AN$

    $$\therefore TS = TA + AS$$
    $$= AN + ZA$$
    $$= ZN$$
    $$= MP$$
    $$= SP$$

And hence, $\lfloor STP = \lfloor SPT$

Again $NG = 2AS = ZS$

$\therefore SG = SN + NG = ZS + SN = MP = SP$

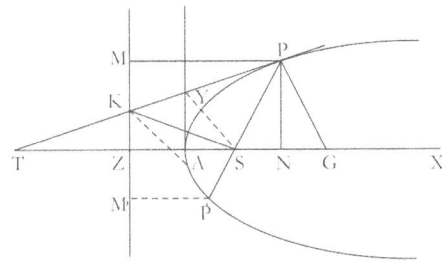

Fig. 91

2. If the tangent at P meet the directrix in k, then KSP is a right angle.

    For $\lfloor SPT = \lfloor PTS = \lfloor KPM$

    Hence, the two triangles KPS and KPM have the two sides KP, PS and the angle KPS equal respectively to the two sides KP, PM and the angle KPM.

    Hence $\lfloor KSP = \lfloor KMP =$ a right angle

    Also $\lfloor SKP = \lfloor MKP$

3. Tangent at the extremities of any focal chord intersect at right angles in the directrix

    For, if PS be produced to meet the curve in P', then, since $\lfloor P'SK$ is a right angle, the tangent at P' meets the directrix in K

    Also $\lfloor MKP = \lfloor SKP$

    And similarly, $\lfloor M'KP' = \lfloor SKP'$

    Hence $\lfloor PKP' = \dfrac{1}{2}\lfloor SKM + \dfrac{1}{2}\lfloor SKM' =$ a right angle

4. If SY be perpendicular to the tangent at P, then Y lies on the tangent at the vertex and $SY^2 = AS \cdot SP$

For the equation to any tangent is

$$y = mx + \frac{a}{m}$$

(1)

The equation to the perpendicular to this line passing through the focus is

$$y = \frac{-1}{m}(x - a)$$

(2)

The lines (1) & (2) meet where

$$mx + \frac{a}{m} = \frac{-1}{m}(x - a) = \frac{-x}{m} + \frac{a}{m}$$

Or $mx - \dfrac{x}{m} = 0$

i.e. where $x = 0$

Hence Y lies on the tangent at the vertex

Also, by geometry $SY^2 = SA \cdot ST = AS \cdot SP$

14. Let us now prove that through any given point (x_1, y_1) there are, in general, two tangents to the parabola.

The equation to any tangent is

$$y = mx + \frac{a}{m} \qquad\qquad (1)$$

If this pass through the fixed point (x_1, y_1), we have

$$y_1 = mx_1 + \frac{a}{m}$$

i.e. $y_1 = \dfrac{m^2 x_1 + a}{m}$

Or $m^2 x_1 - my_1 + a = 0 \qquad\qquad (2)$

For any value of x_1, y_1 we get two values for m

For each value of m (1) given different tangent

15. **Observation**:

Case 1:

The roots of (2) are real and different if $y_1^2 - 4ax_1$ be +ve; i.e. (x_1, y_1) lies without the curve.

Case 2:

They are equal, i.e., the two tangents coalesce into on tangent,

If $y_1^2 - 4ax_1$ be zero. i.e. (x_1, y_1) lie on the curve

Case 3:

The roots are imaginary if $y_1^2 - 4ax_1$ be negative,i.e. point (x_1, y_1) lies within the curve.

16. **Let us obtain an equation to the chord of contact of tangents drawn from a point** (x_1, y_1)

The equation to the tangent at any point Q, whose co-ordinates are x' and y', is $yy' = 2a(x + x')$

If these tangents meet at the point 7, whose co-ordinates are x_1 and y_1 we have

$$y_1 y' = 2a(x_1 + x') \qquad\qquad (1)$$

And $y_1 y'' = 2a(x_1 + x'')$ \qquad\qquad (2)

The equation to QR is then

$$yy_1 = 2a(x + x_1) \qquad\qquad (3)$$

For, since (1) is true, the point (x', y') lies on (3)

Also, since (2) is true, the point (x'', y'') lies on (3)

Hence (3) must be the equation to the straight line joining (x', y') to the point (x'', y'')

i.e. It must be the equation to QR the chord of contact of tangents from the point (x_1, y_1)

17. **Let us obtain an equation of the polar of the point** (x_1, y_1) **with respect to the parabola** $y^2 = 4ax$

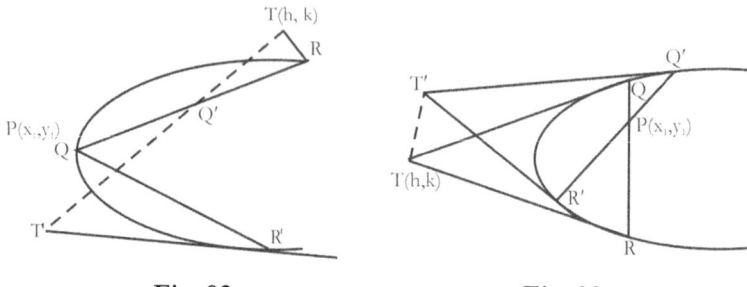

| Fig. 92 | Fig. 93 |

Let Q and R be the points in which any chord drawn through the point P, whose co-ordinates are (x_1, y_1) meets the parabola.

Let the tangents at Q and R meet in the point whose co-ordinates are (h, k)

We required the locus of (h, k)

Since, QR is the chord of contact of tangents from (h, k) its equation is $ky = 2a(x + h)$

Since, this straight line passes through the point (x_1, y_1)

We have $ky_1 = 2a(x_1 + h)$ (1)

Since, the relation (1) true, it follows that the point (h, k) always lies on the straight line

$$yy_1 = 2a(x + x_1) \qquad (2)$$

Hence, (2) is the equation to the polar of (x_1, y_1)

18. **Geometrical construction for the polar of a point** (x_1, y_1)

Let T be the point (x_1, y_1), so that its polar is

$$yy_1 = 2a(x + x_1) \qquad (1)$$

Through T draw a straight line parallel to the axis, its equation is therefore

$$y = y_1 \qquad (2)$$

Let this straight line meet the polar in V and the curve in P

The co-ordinates of V, which is the intersection of (1) and (2), are therefore

$$\frac{y_1^2}{2a} - x_1 \text{ and } y_1 \qquad (3)$$

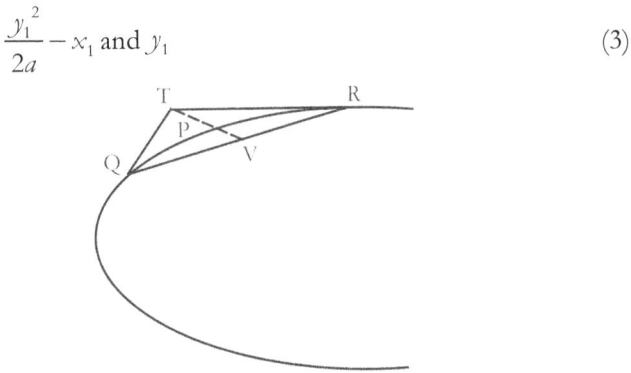

Fig. 94

Also P is the point on the curve whose ordinate is y_1, and whose co-ordinates are therefore

$$\frac{y_1^2}{4a} \text{ and } y_1$$

Since abscissa of $P = \dfrac{\text{abscissa of } T + \text{absciss of } V}{2}$

P is the middle point of T V

Also the tangent at P is

$$yy_1 = 2a\left(x + \frac{y_1^2}{4a}\right), \text{ which is parallel to (1)}$$

Hence, the polar of T is $\parallel^{el}$ to the tangent at P

To draw the polar of T we therefore, draw a line through T parallel to the axis, to meet the curve in P and produce it to V so that

$TP = PV$, a line through V parallel to the tangent at P is then the polar required

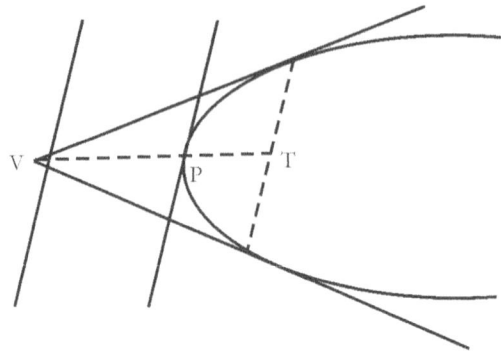

Fig. 95

19. Let us find the pole of a given straight line with respect to the parabola.

Let the given straight line be

$$Ax + By + C = 0$$

If its pole be the point (x_1, y_1), it must be the same straight line as

$$yy_1 = 2a(x + x_1)$$

i.e. $2ax - yy_1 + 2ax_1 = 0$

Since, these straight lines are the same, we have

$$\frac{2a}{A} = \frac{-y_1}{B} = \frac{2ax_1}{C}$$

i.e. $\dfrac{2a}{A} = \dfrac{2ax_1}{C} \Rightarrow x_1 = \dfrac{2a \times C}{A \times 2a} = \dfrac{C}{A}$

And $\dfrac{-y_1}{B} = \dfrac{2a}{A} \Rightarrow y_1 = -\dfrac{2Ba}{A}$

The pole is $\left(\dfrac{C}{A}, \dfrac{-2Ba}{A}\right)$

20. **Let us determine the equation to the pair of tangents that can be drawn to the parabola from the point** (x_1, y_1)

Let (h, k) be any point on either of the tangents drawn from (x_1, y_1). The equation to the line joining (x_1, y_1) to (h, k) is

$$y - y_1 = \frac{k - y_1}{h - x_1}(x - x_1)$$

i.e. $y = y_1 + \dfrac{y - y_1}{h - x_1}(x - x_1)$

$$= \frac{k - y_1}{h - x_1} \cdot x + \frac{hy_1 - kx_1}{h - x_1}$$

If this be a tangent it must be of the form

$$y = mx + \frac{a}{m}$$

So that, $\dfrac{k - y_1}{h - x_1} = m$ and $\dfrac{hy_1 - kx_1}{h - x_1} = \dfrac{a}{m}$

Hence, by multiplication

$$a = \frac{k - y_1}{h - x_1} \cdot \frac{hy_1 - kx_1}{h - x_1}$$

$$a(h - x_1)^2 = (k - y_1)(hy_1 - kx_1)$$

The locus of the point (h, k) is therefore

$$a(x - x_1)^2 = (y - y_1)(xy_1 - yx_1) \tag{1}$$

It will be seen that this equation is the same as

$$(y^2 - 4ax)(y_1^2 - 4ax_1) = \{yy_1 - 2a(x + x_1)\}^2$$

21. **Diameter definition:** The locus of the middle points of a system of parallel chords of a parabola is called a diameter and the chords are called its double ordinates.

22. **The tangents at the ends of any chord meet on the diameters which bisect the chord**

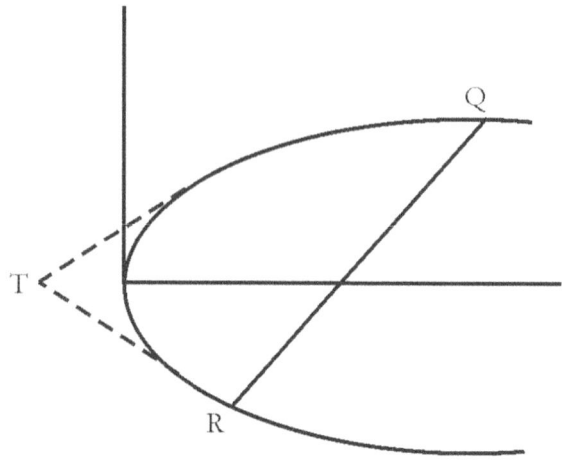

Fig. 96

Let the equation QR be

$$y = mx + c \tag{1}$$

And let the tangent at Q and R meet at the point $T(x_1, y_1)$

The QR is the chord of contacts of tangents drawn from T, and hence its equation is

$$yy_1 = 2a(x + x_1)$$

Comparing this with equation (1), we have

$$\frac{2a}{y_1} = m \text{ , so that } y_1 = \frac{2a}{m}$$

And therefore T lies on the straight line

$$y = \frac{2a}{m}$$

23. **To find the equation to a parabola, the axes being any diameter and the tangent to the parabola at the point where this diameter meets the curve.**

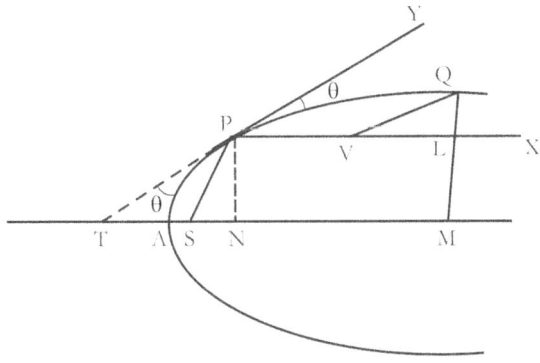

Fig. 97

Let PVX be the diameter and PY the tangent at P meeting the axis in T

Take any point Q on the curve, and draw QM⊥ to the axis meeting the diameter PV in L

Let PV be x and VQ be y

Draw PN⊥ to the axis of the curve, and let

$$\theta = \lfloor YPX = \lfloor PTM$$

Then $4AS \cdot AN = PN^2 = NT^2 \tan^2 \theta = 4AN^2 \cdot \tan^2 \theta$

$$\therefore AN = AS \cdot \cot^2 \theta = a \cot^2 \theta$$

And $PN = \sqrt{4AS \cdot AN} = 2a \cot \theta$

Now $QM^2 = 4AS \cdot AM = 4a AM$ (1)

Also,
$QM = NP + LQ = 2a \cot \theta + VQ \sin \theta = 2a \cot \theta + y \sin \theta$

And $AM = AN + PV + VL = a \cot^2 \theta + x + y \cos \theta$

Substituting these values in (1), we have

$$(2a \cot \theta + y \sin \theta)^2 = 4a(a \cot^2 \theta + x + y \cos \theta)$$

i.e. $y^2 \sin^2 \theta = 4ax$

The required equation is therefore

169

$$y^2 = 4px \tag{2}$$

Where $P = \dfrac{a}{\sin^2\theta} = a(1+\cot^2\theta) = a + AN = SP$

The equation (2) states that

$$QV^2 = 4SP \cdot PV$$

The quantity 4P is called the parameter of the diameter PV. It is equal in length to the chord which is $\|^{d}$ to PY and passes through the focus.

For if $Q'V'R'$ be the chord, $\|^{d}$ to PY and passing through the focus and meeting PV in V', we have

$$PV' = ST = SP = P$$

So that $Q'V'^2 = 4P \cdot PV' = 4P^2$

And hence $Q'R' = 2Q'V' = 4P$

22. Let us express the co-ordinates of any point on the parabola in terms of one variable

It is often convenient to express the co-ordinates of any point on the curve in terms of one variable

It is clear that the values,

$$x = \frac{a}{m^2}, \quad y = \frac{2a}{m}$$

Always satisfy the equation to the curve.

Hence, for all values of m, the point $\left(\dfrac{a}{m^2}, \dfrac{2a}{m}\right)$ lies on the curve.

The equation to the tangent at this point is

$$y = mx + \frac{a}{m}$$

And the normal is $my + x = 2a + \dfrac{a}{m^2}$

The simplest substitution (avoiding both negative signs and fraction) is

$$x = at^2, \text{ and } y = 2at$$

There values satisfy the equation $y^2 = 4ax$

The equations to the tangent and normal at the point $(at^2, 2at)$ are

$$ty = x + at^2$$

$$\text{and } y + tx = 2at + at^3$$

The equation to the straight line joining

$$(at_1^2, 2at_1) \text{ and } (at_2^2, 2at_2)$$

Is $y(t_1 + t_2) = 2x + 2at_1t_2$

The tangent at the points $t_1 + t_2$ are:

$$t_1 y = x + at_1^2$$

$$t_2 y = x + at_2^2$$

The point of intersection of these two tangents is $\{at_1t_2, a(t_1 + t_2)\}$

23. **If the tangents at P and Q meet in T_1 $P \cdot T$ (1) TP and TQ subtend equal angles at the focus S (2) $ST^2 = SP \cdot SQ$ (3) The Δ^{le} SPT & STQ are similar.**

Let P be the point $(at_1^2, 2at_1)$ and Q be the point $(at_2^2, 2at_2)$

So that T is the point $\{at_1t_2, a(t_1 + t_2)\}$

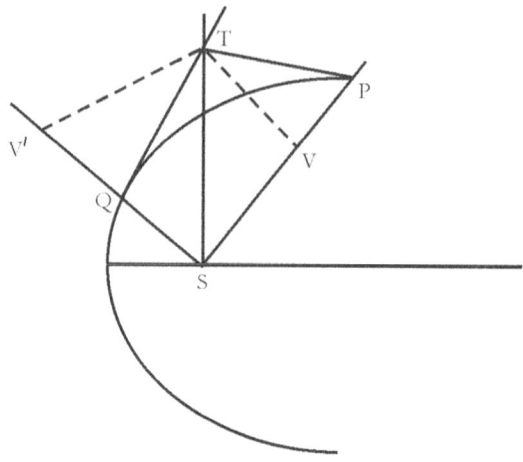

Fig. 98

Case 1:

The equation to SP is

$$y = \frac{2at_1}{at_1^2 - a}(x - a)$$

i.e. $(t_1^2 - 1)y - 2t_1x + 2at_1 = 0$

The $\perp^{lar}$, TV, from T on this straight line

$$= \frac{a(t_1^2 - 1)(t_1 + t_2) - 2t_1 \cdot at_1t_2 + 2at_1}{\sqrt{(t_1^2 - 1)^2 + 4t_1^2}}$$

$$= \frac{a(t_1^3 - t_1^2t_2) + (t_1 - t_2)}{t_1^2 + 1}$$

$$= a(t_1 - t_2)$$

Similarly, TV' has the same numerical value

The angles PST & QST are therefore, equal

Case 2:

We have $SP = a(1 + t_1^2)$ & $SQ = a(1 + t_2^2)$

Also $ST^2 = (at_1t_2 - a)^2 + a^2(t_1 + t_2)^2$

$$= a^2[t_1^2 t_2^2 + t_1^2 + t_2^2 + 1]$$
$$= a^2(1 + c_1^2)(1 + t_2^2)$$

Hence $ST^2 = SP \cdot SQ$

Case 3:

Since $\dfrac{ST}{SP} = \dfrac{SQ}{ST}$ and the angles TSP and TSQ are equal, the Δ^{les} SPT & arc similar so that

$$\lfloor SQT = \lfloor STP \ \& \ \lfloor STQ = \lfloor SPT$$

The area of the Δ^{le} formed by three points on a parabola is twice the area of the Δ^{le} formed by the tangents at these points.

Let the three points on the parabola are

$$(at_1^2, 2at_1), (at_2^2, 2at_2), (at_3^2, 2at_3)$$

The area of the Δ^{le} formed by these points

$$= \frac{1}{2}[at_1^2(2at_2 - 2at_3) + at_2^2(2at_3 - 2at_1)$$
$$+ at_3^2(2at_1 - 2at_2)]$$
$$= -a^2(t_2 - t_3)(t_3 - t_1)(t_1 - t_2)$$

The intersection of the tangents at these points are the points

$$\{at_2 t_3, a(t_2 + t_3)\}, \{at_3 t_1, a(t_3 + t_1)\} \text{ and}$$
$$\{at_1 t_2, a(t_1 + t_2)\}$$

The area of the Δ^{le} formed by there points

$$= \frac{1}{2}\{at_2 t_3(at_3 - at_2) + at_3 t_1(at_1 - at_3) + at_1 t_2(at_2 - at_1)\}$$
$$= \frac{1}{2}a^2(t_2 - t_3)(t_3 - t_1)(t_1 - t_2)$$

24. The circle circumscribing the Δ^k formed by any three tangents to a parabola passes through the focus.

Let P, Q and R be the points at which the tangents are drawn and let their co-ordinates be

$$(at_1^2, 2at_1)(at_2^2, 2at_2) \text{ and } (at_3^2, 2at_3)$$

The tangents at Q and R intersect in the point $\{at_2t_3, a(t_2 + t_3)\}$

Similarly, the other pairs of tangent meet at the points

$$\{at_3t_1, a(t_3 + t_1)\} \text{ and } \{at_1t_2, a(t_1 + t_2)\}$$

Let the equation to the circle be

$$x^2 + y^2 + 2gx + 2fy + c = 0$$
$$(1)$$

Since, it passes through the above three points, we have

$$a^2t_2^2t_3^2 + a^2(t_2 + t_3)^2 + 2gat_2t_3 + 2fa(t_2 + t_3) + c = 0 \quad (2)$$

$$a^2t_3^2 + a^2(t_3 + t_1)^2 + 2gat_3t_1 + 2fa(t_3 + t_1) + c = 0$$
$$(3)$$

And $a^2t_1^2t_2^2 + a^2(t_1 + t_2)^2 + 2gat_1t_2 + 2fa(t_1 + t_2) + c = 0$
$$(4)$$

Subtracting (3) from (2) and dividing by $a(t_2 - t_1)$, we have

$$a\left\{t_3^2(t_1 + t_2) + t_1 + t_2 + 2t_3\right\} + 2gt_3 + 2f = 0$$

Similarly from (3) and (4), we have

$$a\left\{t_1^2(t_2 + t_3) + t_2 + t_3 + 2t_1\right\} + 2gt_1 + 2f = 0$$

From there two equation, we have

$$2g = -a(1 + t_2t_3 + t_3t_1 + t_1t_2) \text{ and}$$

$$2f = -a[t_2 + t_2 + t_3 - t_1t_2t_3]$$

Substituting there values in (2), we obtain

$$C = a^2(t_2t_3 + t_3t_1 + t_1t_2)$$

The equation to the circle is therefore

$$x^2 + y^2 - ax(1 + t_2t_3 + t_3t_1 + t_1t_2)$$
$$-ay(t_1 + t_2 + t_3 - t_1t_2t_3) + a^2(t_2t_3 + t_3t_1 + t_1t_2) = 0$$

which clearly goes through the focus $(a, 0)$

25. **If O be any point on the axis and POP' be any chord passing through O, and if PM and $P'M'$ be the ordinates of P and P', prove that $AM \cdot AM' = AO^2$ and $PM \cdot P'M' = 4a \cdot AO$ for a perable**

Let O be the point (h, O) and let P and P' be the points $(at_1^2, 2at_1)$ and $(at_2^2, 2at_2)$

The equation to PP' is

$$(t_2 + t_1)y - 2x = 2at_1t_2$$

If this pass through the point (h, O), we have

$$-2h = 2at_1t_2$$

$$t_1t_2 = \frac{-h}{a}$$

Hence $AM \cdot AM' = at_1^2 \cdot at_2^2 = a^2\frac{h^2}{a^2} = h^2 = AO^2$

And $PM \cdot PM' = 2at_1 \cdot 2at_2 = 4a^2\left(\frac{-h}{a}\right)$

$$= -4a \cdot AO$$

26. **Corollary: If O be the focus, $AO = a$, and we have** $t_1t_2 = -1$
or $t_2 = \frac{-1}{t_1}$

The points $(at_1^2, 2at_1)$ and $\left(\frac{-a}{t_1^2}, \frac{2a}{t_1}\right)$ are therefore at the ends of focal chord.

Solved Examples

Example 1: Find the equation to the parabola, whose focus is the point $(2,3)$ and whose directrix is the straight line $x-4y+3=0$

$$\text{We have } (x-h)^2 + (y-k)^2 = \left\{ \frac{ax_1 + by_1 + c}{\sqrt{a^2+b^2}} \right\}$$

$$(x-2)^2 + (y-3)^2 = \left[\frac{x-4y+3}{\sqrt{1+16}} \right]^2$$

$$17\left\{ x^2 + y^2 + 4x - 6y + 13 \right\}$$

$$= \left\{ x^2 + 16y^2 + 9 - 8xy + 6x - 24y \right\}$$

i.e. $16x^2 + 16y^2 + 8xy - 74x - 78y + 212 = 0$

Example 2 : Find the vertex, axis, focus and latus rectum of the parabola $4y^2 - 12x - 20y + 67 = 0$

The equation can be written

$$4y^2 - 20y = -12x - 67 \div 4$$

$$y^2 - 5y = -3x - \frac{67}{4}$$

$$y^2 - 2 \times \frac{5}{2} y + \left(\frac{5}{2}\right)^2 = -3x - \frac{67}{4} + \left(\frac{5}{2}\right)^2$$

$$\left(y - \frac{5}{2}\right)^2 = -3x - \left(\frac{67}{4} - \frac{25}{4}\right)$$

$$= -3x - \frac{42}{4}$$

$$\left(y - \frac{5}{2}\right)^2 = -3\left(x + \frac{7}{2}\right)$$

Transform this equation to the point $\left(\dfrac{-7}{2}, \dfrac{5}{2}\right)$ and it becomes

$Y^2 = -3x$ where $Y = y\dfrac{-5}{2}, X = x + \dfrac{7}{2}$ which represents a parabola, whose axis is the axis of x and whose eccentricity is turned towards the negative end of this axis. Also its latus rectum is 3.

Referred to the original axes the vertex is the point $\left(\dfrac{-7}{2}, \dfrac{5}{2}\right)$ the

axis is $y = \dfrac{5}{2}$ and the found is the point $\left(\dfrac{-7}{2} - \dfrac{3}{4}, \dfrac{5}{2}\right)$

i.e. $\left(\dfrac{-14-3}{4}, \dfrac{5}{2}\right)$ or $\left(\dfrac{-17}{4}, \dfrac{5}{2}\right)$

Example 3: The equation to the tangent at the point $(2, -4)$ of the parabola $y^2 = 8x$ is $y(-4) = 4(x+2)$

The equation $y(-4) = 4(x+2)$ can be rewritten as

$\qquad -4y = 4x + 8$

$\qquad \text{Or } x + y + 2 = 0$

The equation to the tangent at the point $\left(\dfrac{a}{m^2}, \dfrac{2a}{m}\right)$ of the parabola $y^2 = 4ax$ is

$$y \cdot \dfrac{2a}{m} = 2a\left(x + \dfrac{a}{m^2}\right) \div 2a$$

$$\dfrac{y}{m} = \dfrac{xm^2 + a}{m^2}$$

$\text{Or } y = mx + \dfrac{a}{m}$

Also, it is the tangent at the point (x', y')

i.e. $\left(\dfrac{a}{m^2}, \dfrac{2a}{m}\right)$

Example 4: If a chord which is normal to the parabola at one end subtend a right angle at the vertex, prove that it is inclined at an angle $\tan^{-1}\sqrt{2}$ to the axis.

The equation to any chord which is normal, is

$$y = mx - 2am - am^3,$$

i.e. $mx - y = 2am + am^3$

The parabola is $y^2 = 4ax$

The straight line joining the origin to the intersections of these two are, therefore, given by the equation

$$y^2(2am + am^3) - 4ax(mx - y) = 0$$

If these be at right angles, then

$$2am + am^3 - 4am = 0$$

$$am^3 - 2am = 0 \div am$$

$$m^2 - 2 = 0$$

$$\therefore m = \pm\sqrt{2}$$

Example 5: Find the locus of the intersection of tangents to the parabola $y^2 = 4ax$, the angle between being always a given angle α.

The straight line $y = mx + \dfrac{a}{m}$ is always a tangent to the parabola

If it pass through the point T(h, k), we have

$$m^2h - mk + a = 0 \tag{1}$$

If m_1 and m_2 be the roots of this equation, we have

$$m_1 + m_2 = \dfrac{k}{h} \tag{2}$$

And $m_1 m_2 = \dfrac{a}{b}$ (3)

And the equations to TP and PQ are, then

$$y = m_1 x + \dfrac{a}{m_1} \text{ and } y = m_2 x + \dfrac{a}{m_2}$$

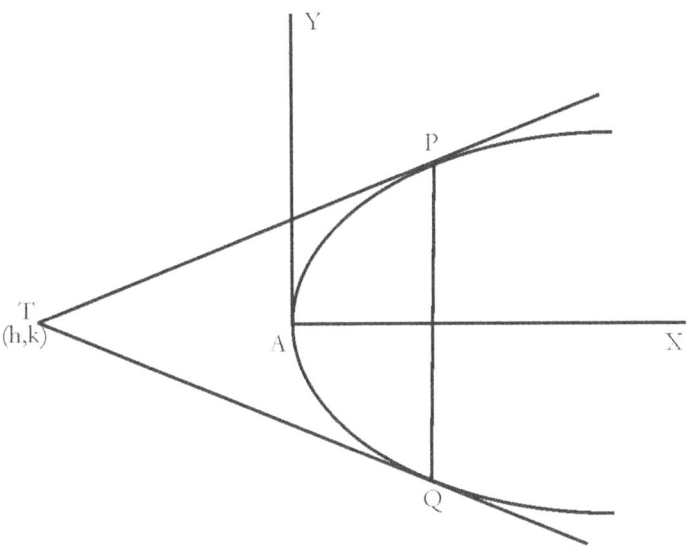

Fig. 99

Hence, we have

$$\tan \alpha = \dfrac{m_1 - m_2}{1 + m_1 m_2} = \dfrac{\sqrt{(m_1 + m_2)^2 - 4 m_1 m_2}}{1 + m_1 m_2}$$

$$= \dfrac{\sqrt{\left(\dfrac{k}{b}\right)^2 - 4\left(\dfrac{a}{b}\right)}}{1 + \dfrac{a}{b}}$$

$$= \frac{\sqrt{\dfrac{k^2 - 4ab}{b^2}}}{\dfrac{a+b}{b}} = \frac{\sqrt{k^2 - 4ab}}{a+b} \quad (\because \text{by (2) \& (3)})$$

$$\therefore (\tan\alpha)(a+b) = \sqrt{k^2 - 4ab}$$

Squaring both sides, we get

$$k^2 - 4ab = (a+b)^2 \tan^2\alpha$$

Hence the co-ordinates of the point T always satisfy the equation $y^2 - 4ax = (a+x)^2 \tan^2\alpha$

Example 6: Prove that the locus of the poles of chords which are normal to the parabola $y^2 = 4ax$ is the curve $y^2(x+2a) + 4a^3 = 0$

Let PQ be a chord which is normal at p. Its equation is then

$$y = mx - 2am - am^3$$
$$(1)$$

Let the tangents at P and Q intersect in T, whose co-ordinates are h and k, so that we require the locus of T

Since, PQ is the polar of the point (h, k) its equation is

$$yk = 2a(x+h)$$
$$(2)$$

Now the equations (1) and (2) represent the same straight line, so that they must be equivalent

Hence $m = \dfrac{2a}{k}$ and $-2am - am^3 = \dfrac{2ah}{k}$

Substituting $m = \dfrac{2a}{k}$ in $-2am - am^3 = \dfrac{2ab}{k}$, we get

$$-2a\left(\frac{2a}{k}\right) - a\left(\frac{2a}{k}\right)^3 = \frac{2ab}{k}$$

Or $\dfrac{-4a^2}{k} - \dfrac{8a^4}{k^3} = \dfrac{2ah}{k}$

i.e. $\dfrac{k^2(4a^2+2ab)-8a^4}{k^3}=0$

Or $2ak^2(2a+b)+2a\cdot 4a^3=0$

$\div 2a$, we get

$k^2(b+2a)+4a^3=0$

The locus of the point T is therefore

$y^2(x+2a)+4a^3=0$

Example 7: Find the locus of the middle points of chords of a parabola which subtend a right angle at the vertex, and prove that these chords all pass through a fixed point on the axis of the curve.

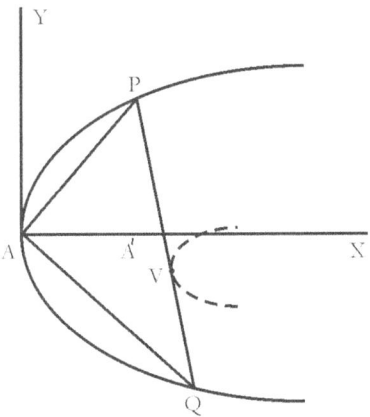

Fig. 100

Let PQ be any chord, and let its equation be

$$y=mx+c$$

$$(1)$$

The lines joining the vertex with the points of intersection of this straight line with the parabola

$$y^2=4ax$$

$$(2)$$

Are given by the equation

$$y^2 c = 4ax(y - mx)$$

These straight lines are at right angles if,

$$c + 4am = 0 \Rightarrow c = -4am$$

Substituting this value of c in (1), the equation to PQ is

$$y = mx - 4am$$

$$y = m(x - 4a)$$
$$(3)$$

This straight line cuts the axis of x at a constant distance 4a from the vertex.

i.e. $AA' = 4a$

if the middle point of PQ be (h, k), we have

$$k = \frac{2a}{m}$$
$$(4)$$

Also the point (h, k) lies on (3), so that we have

$$k = m(h - 4a)$$
$$(5)$$

If between (4) & (5) we eliminate m_1 we have

$$m = \frac{2a}{k}$$

$$\therefore k = \frac{2a}{k}(h - 4a)$$

Or $k^2 = 2a(h - 4a)$

So that (h, k) always lies on the parabola

$$y^2 = 2a(x - 4a)$$

Example 8: Prove that, ingeneral, three normals can be drawn from any point to the parabola and that the algebraic sum of the ordinates of the feet of these normal is zero

The straight line

$$y = mx - 2am - am^3 \qquad (1)$$

Is normal to the parabola at the point whose co-ordinates are

$$am^2 \text{ and } -2am \qquad (2)$$

If this normal passes through the fixed point O, whose co-ordinates are h and k, we have

$$k = mh - 2am - am^3$$

i.e. $am^3 + (2a - h)m + k = 0 \qquad (3)$

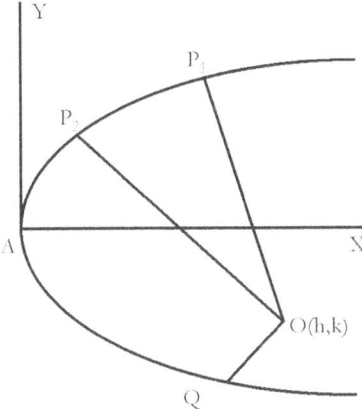

Fig. 101

This equation being of the third degree, has three roots real, or imaginary. Corresponding to each of these roots, we have, co substitution in (1), the equation to a normal which passes through the point O.

If m_1, m_2 and m_3 are the roots of the equation (3)

If the ordinates of the feet of these normal be $y_1, y_2 \& y_3$, we have by (2)

$$y_1 + y_2 + y_3 = -2a(m_1 + m_2 + m_3) = 0$$

Example 9: Find the locus of a point which is such that (a) two of the normal drawn from it to the parabola are at right angles, (b) the three normal through it cut the axis in points whose distance from the vertex are in arithmetical progression.

Any normal is $y = mx - 2am - am^3$, and this passes through the point (h, k), if

$$am^3 + (2a - h)m + k = 0 \qquad (1)$$

If then m_1, m_2 and m_3 be the roots, we have

$$m_1 + m_2 + m_3 = 0 \qquad (2)$$

$$m_2 m_3 + m_3 m_1 + m_1 m_2 = \frac{2a - h}{a} \qquad (3)$$

And $m_1 m_2 m_3 = \frac{-k}{a} \qquad (4)$

(A) If two of the normal say m_1 and m_2, be at right angles, we have $m_1 m_2 = -1$ and hence, from (4), $m_3 = \frac{k}{a}$.

The quantity $\frac{k}{a}$ is therefore, a root of (1) and hence, by substitution, we have

$$\frac{k^3}{a^2} + (2a - h)\frac{k}{a} + k = 0$$

i.e. $k^2 = a(h - 3a)$

The locus of the point (h, k) is therefore the parabola $y^2 = a(x - 3a)$ whose vertex is the point $(3a, 0)$ and whose latus rectum is one-quarter that of the given parabola.

(B) The normal $y = mx - 2am - am^3$ meets the axis of x at a point whose distance from the vertex is $2a + am^2$.

The conditions of the question, then gives

$$(2a + am_1^2) + (2a + am_3^2) = 2(2a + am_2^2)$$

i.e. $m_1^2 + m_3^2 = 2m_2^2 \qquad (5)$

If we eliminate $m_1 m_2$ and m_3 from the equations (2), (3) (4) & (5) we shall have a relation between h & k

From (2) (3), we have

$$\frac{2a-h}{a} = m_1 m_3 + m_2(m_1 + m_3) = m_1 m_3 - m_2^2 \qquad (6)$$

Also (5) and (2) gives

$$2m_2^2 = (m_1 + m_3)^2 - 2m_1 m_3$$
$$= m_2^2 - 2m_1 m_3$$

i.e. $m_2^2 + 2m_1 m_3 = 0$ \qquad (7)

Solving (6) & (7), we have

$$m_1 m_3 = \frac{2a-h}{3a} \text{ and } m_2^2 = -2\left(\frac{2a-h}{3a}\right)$$

Substituting these values in (4), we have

$$\frac{2a-h}{3a}\sqrt{-2 \cdot \frac{2a-h}{3a}} = \frac{-k}{a}$$

i.e. $27ak^2 = 2(h-2a)^3$

The required locus is

$$27ay^2 = 2(x-2a)^3$$

11. The Ellipse

1. **Ellipse:**The ellipse is a conic section in which the eccentricity is less than unity.
2. Let us derive the equation to an ellipse.

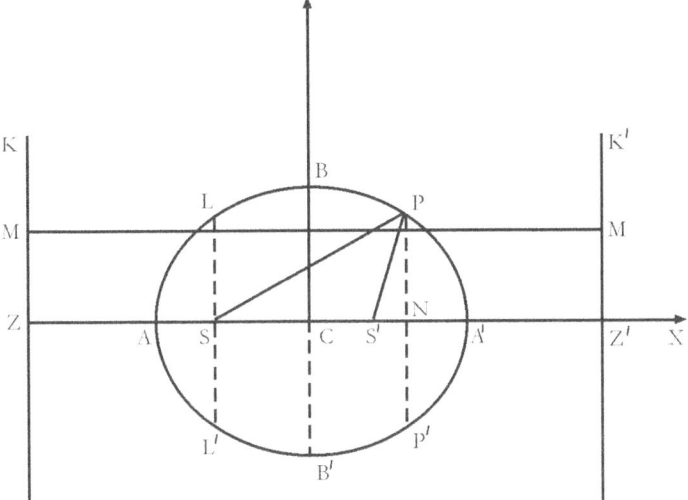

Fig. 102

Let ZK be the directrix, S the focus, and let SZ be $\perp$ to the directix.

There will be a point A on SZ, such that

$$SA = e \cdot AZ$$
$$(1)$$

Since, $e < 1$, there will be another point A', in ZS produced, such that

$$SA' = eA'Z$$
$$(2)$$

Let the length AA' be called $2a$, and let D be the middle point of AA', adding equations (1) and (2), we have

$$SA + SA' = e(AZ + A'Z)$$

$$AA' = e(AZ + A'Z)$$

$$= e(CZ - CA + CA' + CZ) \; (\because CA = CA')$$

$$= 2eCZ$$

$$\therefore CZ = \frac{2a}{2e}$$

$$CZ = \frac{a}{e}$$

(3)

Subtracting equations (1) and 92), we have

$$SA - SA' = e(AZ - A'Z)$$

$$= eAA'$$

$$CA - CS - (CS + CA') = -e(2a)$$

$$-2CS = -2ae$$

$$CS = \frac{2ae}{2}$$

$$CS = ae$$

(4)

Let C be the origin, CA' the axis of x, and a line through C $\perp$ to AA' the axis of y

Let P be any point on the curve, whose co-ordinates are x and y, and let PM be the $\perp$ upon the directrix, and PN the $\perp$ upon AA'

The focus S is the point $(-ae, 0)$

The relation $SP^2 = e^2 PM^2 = e^2 ZN^2$, then gives

$$SP = \sqrt{[x - (-ae)]^2 + (y - 0)^2}$$

$$SP^2 = (x + ae)^2 + y^2$$

$$\therefore (x+ae)^2 + y^2 = e^2\left(x+\frac{a}{e}\right)^2$$

$$x^2 + a^2 e^2 + 2xae + y^2 = e^2\left(x^2 + \frac{a^2}{e^2} + \frac{2a}{e}x\right)$$

$$x^2 + a^2 e^2 + 2xae + y^2 + e^2 x^2 + a^2 + 2aex$$

Or $x^2(1-e^2) + y^2 = a^2(1-e^2)$

$$\div a^2(1-e^2) \text{ given}$$

$$\frac{x^2}{a^2} + \frac{y^2}{a^2(1-e^2)} = 1$$

$$(5)$$

If in this equation we put $x = 0$, we have

$$y^2 = a^2(1-e^2)$$

Or $y = \pm a\sqrt{1-e^2}$

Thus the curve meets the axis of y in two points B and B', lying on opposite sides of C, such that

$$B'C = CB = a\sqrt{1-e^2}$$

i.e. $CB^2 = CA^2 - CS^2$

Let the length CB be called b, so that

$$b = a\sqrt{1-e^2}$$

Thus equation (5), then becomes

$$\frac{x^2}{a^2} + \frac{y^2}{b^2} = 1$$

$$(6)$$

3. **Definition major and minor axis:** The points A and A' are called the vertices of the curve, AA' is called the major axis and BB' is the minor axis. Also C is called center.

4. **Observations:**

Case 1:

If the focus is $S(-ae, 0)$, then equation of ellipse is

$$\frac{(x-ae)^2}{a^2} + \frac{y^2}{b^2} = 1$$

Case 2:

The equation referred to A as origin, and AX and a $\perp$ line as axes is

$$\frac{(x-a)^2}{a^2} + \frac{y^2}{b^2} = 1$$

$$\frac{x^2 + a^2 - 2ax}{a^2} + \frac{y^2}{b^2} = 1$$

$$\text{Or } \frac{x^2}{a^2} + \frac{y^2}{b} - \frac{2x}{a} = 0$$

Case 3:

The equation referred to zx and zk as axes is, since

$$CZ = \frac{-a}{e}$$

$$\frac{\left(x - \frac{a}{e}\right)^2}{a^2} + \frac{y^2}{b^2} = 1$$

Case 4:

Generally, the equation to the ellipse, whose focus is the point (f, g) whose directrix is $Ax + By + C = 0$, and whose eccentricity is e, is

$$(x-f)^2 + (y-g)^2 = e^2 \frac{(Ax + By + C)^2}{A^2 + B^2}$$

The sum of the focal distances of any point on the curve is equal to the major axis.

We have $SP = ePM$ and $S'P = ePM'$

Hence $SP + S'P = e(PM + PM')$

$$= eMM'$$
$$= eZZ'$$
$$= 2eCZ$$
$$= 2a$$

The major axis

Also $SP = ePM = eNZ = eCZ + eCN = a + ex'$

$$S'P = ePM' = eNZ' = eCZ' - eCN = a - ex'$$

5. Latus – rectum of an ellipse

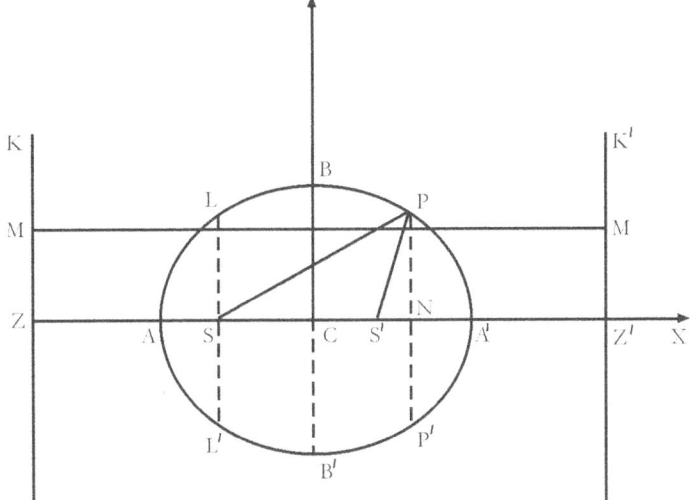

Fig 103

Let LSL' be the double ordinate of the curve which passes through the focus S. By the definition of the curve, the semi-latus rectum SL

= e times the distance of L from the directrix

$$= eSZ = e(CZ - CS)$$

$$= eCZ - eCS$$

$$(\because CZ = \frac{a}{e}$$

$$CS = ae$$

$$= e \cdot \frac{a}{e} - e \cdot ae$$

$$= a - ae^2$$

$$= a(1 - e^2)$$

$$= \frac{b^2}{a}$$

The quantity $\dfrac{x'^2}{a^2} - \dfrac{y'^2}{b^2} - 1$ is negative, zero, or positive accord-

ing as the point (x', y') lies within, upon, or without the ellipse.

Let Q be the point (x', y'), and let the ordinate QN through Q meet the curve in P, so that, by equation (6)

$$\frac{PN^2}{b^2} = 1 - \frac{x'^2}{a^2}$$

i.e. eq (6) is $\dfrac{x^2}{a^2} + \dfrac{y^2}{b^2} = 1$

if Q be within the curve, then y'

i.e. QN < PN, so that

$$\frac{y'^2}{b^2} < \frac{PN^2}{b^2}$$

i.e. $< 1 - \dfrac{x'^2}{a^2}$

Hence, in this case

$$\frac{x'^2}{a^2} + \frac{y'^2}{b^2} < 1$$

i.e. $\dfrac{x'^2}{a^2} + \dfrac{y'^2}{b^2} - 1$ is negative

Similarly, if Q' be without the curve, $y' > PN$, and then

$$\frac{x'^2}{a^2} + \frac{y'^2}{b^2} - 1 \text{ is positive}$$

6. **Let us determine the length of a radius vector from the centre drawn in a given ellipse.**

We have equation to ellipse $\dfrac{x^2}{a^2} + \dfrac{y^2}{b^2} = 1$

Put $x = r\cos\theta$, $y = r\sin\theta$, then

$$\frac{(r\cos\theta)^2}{a^2} + \frac{(r\sin\theta)^2}{b^2} = 1$$

$$\frac{b^2 r^2 \cos^2\theta + a^2 r^2 \sin^2\theta}{a^2 b^2} = 1$$

$$r^2(b^2\cos^2\theta + a^2\sin^2\theta) = a^2 b^2$$

$$r^2 = \frac{a^2 b^2}{b^2\cos^2\theta + a^2\sin^2\theta}$$

This given the radius vector drawn at any inclination θ to the axis
If $\theta = 0$ then

$$r^2 = \frac{a^2 b^2}{b^2\cos^2 0 + a^2\sin^2 0}$$

$$= \frac{a^2 b^2}{b^2}$$

$$r^2 = a^2$$

The greatest value of r when $\theta = 0$ is $r = a$
If $\theta = 90$, then

$$r^2 = \frac{a^2 b^2}{b^2\cos^2 90 + a^2\sin^2 90}$$

$$= \frac{a^2 b^2}{b^2(0) + a^2(1)}$$

$$= \frac{a^2 b^2}{a^2}$$

$$= b^2$$

The least value of r when $\theta = 90$ is r = b

Also for each value of θ, we have two equal and opposite values of r, so that any line through the centre meets the curve in two points equidistance from it.

7. **Auxiliary circle:** The circle which is described on the major axis, AA' of an ellipse as diameter, is called the auxiliary circle of the ellipse.

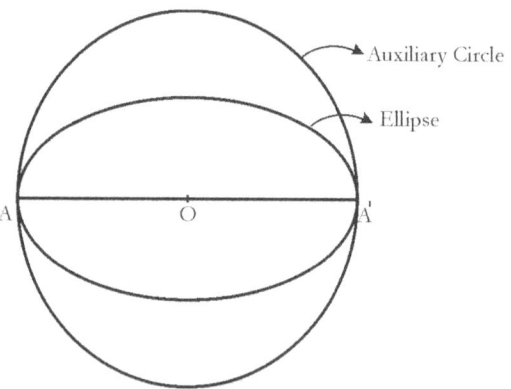

Fig. 104

8. **Eccentric Angle**: The eccentric angle of any point P on the ellipse is the angle NCQ made with the major axis by the straight line CQ joining the centre C to the point Q on the auxiliary circle which corresponds called ϕ.

We have $CN = CQ \cdot \cos\phi = a \cos\phi$

And $NQ = CQ \cdot \sin\phi = a \sin\phi$

Hence, we have $NP = \frac{b}{a} NQ = b \sin\phi$

The co-ordinates of any point P on the ellipse are

Therefore $a\cos\phi$ and $b\sin\phi$

Since P is known when ϕ is given, it is also called "the point ϕ"

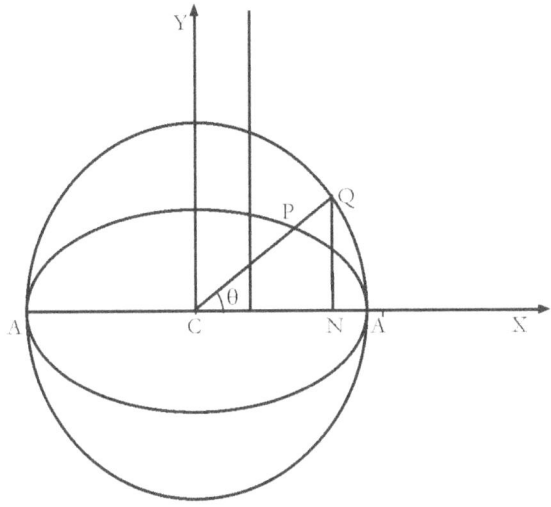

Fig. 105

9. **Let us now obtain the equation of the straight line joining two points on the ellipse whose eccentric angles are given.**

Let the eccentric angles of the two points, P and P', be ϕ & ϕ', so that the points have as co-ordinates

$$(a\cos\phi, b\sin\phi) \text{ and } (a\cos\phi', b\sin\phi')$$

The equation of the straight line joining them is

$$y - b\sin\phi = \frac{b\sin\phi' - b\sin\phi}{a\cos\phi' - a\cos\phi}(x - a\cos\phi)$$

$$= \frac{b(\sin\phi' - \sin\phi)}{a(\cos\phi' - \cos\phi)}(x - a\cos\phi)$$

$$= \frac{b \cdot 2 \cos\left(\frac{\phi + \phi'}{2}\right) \sin\left(\frac{\phi' - \phi}{2}\right)}{a 2 \sin\left(\frac{\phi + \phi'}{2}\right) \sin\left(\frac{\phi - \phi'}{2}\right)} (x - a \cos\phi)$$

$$= \frac{-b \cos\left(\frac{\phi + \phi'}{2}\right) \sin\left(\frac{\phi - \phi'}{2}\right)}{a \sin\left(\frac{\phi + \phi'}{2}\right) \sin\left(\frac{\phi - \phi'}{2}\right)} (x - a \cos\phi)$$

$$y - b \sin\phi = \frac{-b}{a} \frac{\cos\left(\frac{\phi + \phi'}{2}\right)}{\sin\left(\frac{\phi + \phi'}{2}\right)} (x - a \cos\phi)$$

$$y - b \sin\phi = \frac{-b}{a} \frac{\cos\left(\frac{\phi + \phi'}{2}\right)}{\sin\left(\frac{\phi + \phi'}{2}\right)} (x - a \cos\phi)$$

$$(y - b \sin\phi) \sin\left(\frac{\phi + \phi'}{2}\right) = \frac{-b}{a} \cos\left(\frac{\phi + \phi'}{2}\right)(x - a \cos\phi)$$

$$y \sin\frac{\phi + \phi'}{2} - b \sin\phi \sin\left(\frac{\phi + \phi'}{2}\right) = \frac{-b}{a} \cdot x \cos\left(\frac{\phi + \phi'}{2}\right)$$

$$+ b \cos\left(\frac{\phi + \phi'}{2}\right) \cos\phi$$

$$\text{Or} \frac{x}{a} \cos\left(\frac{\phi + \phi'}{2}\right) + \frac{y}{b} \sin\left(\frac{\phi + \phi'}{2}\right)$$

$$= \cos\phi \cos\left(\frac{\phi + \phi'}{2}\right) + \sin\phi \cdot \sin\left(\frac{\phi + \phi'}{2}\right)$$

$$= \cos\left[\phi - \frac{\phi + \phi'}{2}\right]$$

$$= \cos\left[\frac{\phi - \phi'}{2}\right]$$

$$(1)$$

This is the required equation

10. **Corollary**: The points on the auxiliary circle, corresponding to P and P', hence the co-ordinates $(a\cos\phi, a\sin\phi)$ and $(a\cos\phi', a\sin\phi')$

The equation to the line joining them is therefore

$$\frac{x}{a}\cos\left(\frac{\phi + \phi'}{2}\right) + \frac{y}{a}\sin\left(\frac{\phi + \phi'}{2}\right) = \cos\left(\frac{\phi - \phi'}{2}\right)$$

The straight line and equation (1) clearly make the same intercept on the major axis.

Hence, the straight line joining any two points on an ellipse, and the straight line joining the corresponding points on the auxiliary circle; meet the major axis in the same point.

11. **Let us now determine the intersections of any straight line with the ellipse** $\dfrac{x^2}{a^2} + \dfrac{y^2}{b^2} = 1$

The general equation of an ellipse is $\dfrac{x^2}{a^2} + \dfrac{y^2}{b^2} = 1$ (1)

Let the equation of the straight line be

$$y = mx + c \qquad (2)$$

The co-ordinates of the points of intersection of (1) & (2) satisfy both equations and therefore, obtained by solving them as simultaneous equations

Substituting for y in (1) from (2), the abscissa of the points of intersection are given by the equation

$$\frac{x^2}{a^2} + \frac{(mx + c)^2}{b^2} = 1$$

$$\frac{b^2 x^2 + a^2(m^2 x^2 + c^2 + 2mcx)}{a^2 b^2} = 1$$

Or $x^2(a^2m^2 + b^2) + 2a^2mcx + a^2(c^2 - b^2) = 0$ (3)

This is a quadratic equation and hence has two roots, real co-incident, or imaginary.

Also, corresponding to each value of x, we have eq (2) one value of y.

The straight line therefore meets the curve in two points real, coincident or imaginary.

The roots of the equation (3) are real, coincident, or imaginary according as

$$(2a^2mc)^2 - 4(b^2 + a^2m^2) \times a^2(c^2 - b^2) \text{ is positive, zero,}$$
or negative

According as $b^2(b^2 + a^2m^2) - b^2c^2$ is +ve, zero, or negative

i.e. according as c^2 is $<=$ or $>a^2m^2 + b^2$

12. Let us now obtain the length of the chord intercepted by the ellipse, on the straight line $y = mx + c$

We have $x_1 + x_2 = \dfrac{-2a^2mc}{a^2m^2 + b^2}$ and $x_1x_2 = \dfrac{a^2(c^2 - b^2)}{a^2m^2 + b^2}$

$$(x_1 - x_2)^2 = (x_1 + x_2)^2 - 4x_1x_2$$

$$= \left(\frac{-2a^2mc}{a^2m^2 + b^2}\right)^2 - 4\left[\frac{a^2(c^2 - b^2)}{a^2m^2 + b^2}\right]$$

$$= \frac{4a^4m^2c^2}{(a^2m^2 + b^2)^2} - \frac{4a^2(c^2 - b^2)}{a^2m^2 + b^2}$$

$$= \frac{4a^4m^2c^2 - 4a^2(c^2 - b^2)(a^2m^2 + b^2)}{(a^2m^2 + b^2)^2}$$

So $x_1 - x_2 = \dfrac{2ab\sqrt{a^2m^2 + b^2 - c^2}}{a^2m^2 + b^2}$

$y_1 - y_2 = m(x_1 - x_2)$

The length of the required chord, therefore

$$= \sqrt{(x_1 - x_2)^2 + (y_1 - y_2)^2}$$

$$= x_1 - x_2 \sqrt{1 + m^2}$$

$$= \frac{2ab\sqrt{1 + m^2}\sqrt{a^2 m^2 + b^2 - c^2}}{a^2 m^2 + b^2}$$

12. **Let us derive the equation to the tangent at any point** (x', y') **of the ellipse.**

Let P and Q be two points on the ellipse, whose co-ordinates are (x', y') and (x'', y'')

The equation to the straight line PQ is

$$y - y' = \frac{y'' - y'}{x'' - x'}(x - x') \quad (1)$$

Since, both P and Q lie on the ellipse, we have

$$\frac{x'^2}{a^2} + \frac{y'^2}{b^2} = 1 \quad\quad (2)$$

And $\dfrac{x''^2}{a^2} - \dfrac{y''^2}{b^2} = 1$ $\quad\quad (3)$

Hence, by sub

$$\frac{x''^2}{a^2} - \frac{x'^2}{a^2} + \frac{y''^2}{b^2} - \frac{y'^2}{b^2} = 0$$

$$\frac{(x'' + x')(x'' - x')}{a^2} + \frac{(y'' + y')(y'' - y')}{b^2} = 0$$

Or $\dfrac{(x'' + x')(x'' - x')}{a^2} = \dfrac{-(y'' + y')(y'' - y')}{b^2}$

Or $\dfrac{y'' - y'}{x'' - x'} = \dfrac{-b^2}{a^2}\dfrac{(x'' + x')}{(y'' + y')}$

On substituting in equation (1), the equation to any secant PQ becomes

$$y - y' = \frac{-b^2(x'' + x')}{a^2(y'' + y')}(x - x')$$

To obtain the equation to the tangent we take Q indefinitely close to P, and hence, in the limit, we put $x'' = x'$ and $y'' = y'$

The equation (4) then becomes

$$y - y' = \frac{-b^2 x'}{a^2 y'}(x - x')$$

i.e. $a^2 y'(y - y') = -b^2 x'(x - x')$

$$a^2 yy' - a^2 y'^2 = -b^2 xx' + b^2 x'^2$$

$$b^2 xx' + a^2 yy' = b^2 x'^2 + a^2 y'^2 \div a^2 b^2$$

$$\frac{xx'}{a^2} + \frac{yy'}{b^2} = \frac{x'^2}{a^2} + \frac{y'^2}{b^2} = 1 \text{ (From (2))}$$

The required eq is $\dfrac{xx'}{a^2} + \dfrac{yy'}{b^2} = 1$

13. **We will now derive the equation to a tangent in terms of the tangent of its inclination to the major axis**

The straight line $y = mx + c$ (1)

Meets the ellipse in points whose abscissa are given by

$$x^2(b^2 + a^2 m^2) + 2mca^2 x + a^2(c^2 - b^2) = 0$$

And the roots of this equation are coincident if

$$c = \sqrt{a^2 m^2 + b^2}$$

In this case the straight line (1) is a tangent, and it becomes

$$y = mx + \sqrt{a^2 m^2 + b^2}$$

(2)

This is the required equation

Similarly if the straight line $x \cos \alpha + y \sin \alpha = P$ touches the ellipse if

$$P^2 = a^2 \cos^2 \alpha + b^2 \sin^2 \alpha$$

The straight line $lx + my = n$ touched the ellipse, if

$$a^2 l^2 + b^2 m^2 = n^2$$

14. **We will derive now equation to the normal at the point** (x', y').

The required normal is the straight line which passes through the point (x', y') and is $\perp$ to the tangent

i.e. to the straight line

$$y = \frac{-b^2 x'}{a^2 y'} x + \frac{b^2}{y'}$$

Its equation is therefore

$$y - y' = m(x - x')$$

Where $m \left(\dfrac{-b^2 x'}{a^2 y'} \right) = -1$

Or $m = \dfrac{a^2 y'}{b^2 x'}$

The equation to the normal is therefore

$$y - y' = \frac{a^2 y'}{b^2 x'}(x - x')$$

Or $\dfrac{x - x'}{\dfrac{x'}{a^2}} = \dfrac{y - y'}{\dfrac{y'}{b^2}}$

15. **We will derive the equation to the normal at the point whose eccentric angle is** ϕ

We have $x' = a \cos \phi$ and $y' = b \sin \phi$

It becomes $\dfrac{x - a \cos \phi}{\dfrac{\cos \phi}{a}} = \dfrac{y - b \sin \phi}{\dfrac{\sin \phi}{b}}$

$$\frac{a(x - a\cos\phi)}{\cos\phi} = \frac{b(y - b\sin\phi)}{\sin\phi}$$

i.e. $\dfrac{ax}{\cos\phi} - a^2 = \dfrac{by}{\sin\phi} - b^2$

The required normal is therefore

$$ax\sec\phi - by\cosec\phi = a^2 - b^2$$

16. Let us now represent the equation to the normal in the form $y = mx + c$

The equation to the normal at (x', y') is

$$y = \frac{a^2 y'}{b^2 x'} x - y'\left(\frac{a^2}{b^2} - 1\right)$$

Let $\dfrac{a^2 y'}{b^2 x'} = m$, so that $\dfrac{x'}{a} = \dfrac{ay'}{b^2 m}$

Hence, since (x', y') satisfies the relation $\dfrac{x'^2}{a^2} + \dfrac{y'^2}{b^2} = 1$

We obtain $y' = \dfrac{b^2 m}{\sqrt{a^2 + b^2 m^2}}$

17. Let us determine the length of the sub tangent and sub normal.

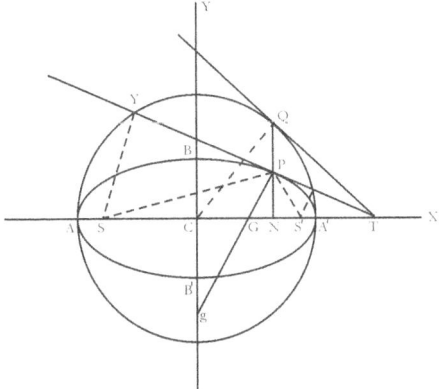

Fig. 106

Let the tangent and normal at P, the point (x', y'), meet the axis in T and G respectively, and let PN be the ordinate of P. The equation to the tangent at P is

$$\frac{xx'}{a^2} + \frac{yy'}{b^2} = 1$$

(1)

To find where the straight line meets the axis, we put $y = 0$

And $x = \dfrac{a^2}{x'}$

i.e. $CT = \dfrac{a^2}{CN}$

$$CT \cdot CN = a^2 = CA^2$$

(2)

Hence, the sub tangent $NT = CT - CN$

$$= \frac{a^2}{x'} - x'$$

$$= \frac{a^2 - x'^2}{x'}$$

The equation to the normal is

$$\frac{x - x'}{\dfrac{x'}{a^2}} = \frac{y - y'}{\dfrac{y'}{b^2}}$$

To find where it meets the axis, we put $y = 0$, and we have

$$\frac{x - x'}{\dfrac{x'}{a^2}} = \frac{-y'}{\dfrac{y'}{b^2}} = -b^2$$

i.e. $CG = x = x' - \dfrac{b^2}{a^2}x' = \dfrac{a^2 - b^2}{a^2}x' = e^2 x' = e^2 \cdot CN$

(3)

Hence, the sub normal $NG = CN - CG = (1 - e^2)CN$

i.e. $NG : NC :: 1 - e^2 : 1$

$:: b^2 : a^2$

18. **Corollary:** If the tangent meet the minor axis in t, and P n be $\perp$ to it, we may, similarly, prove that

$ct \cdot cn = b^2$

19. Let us now enumerate a few properties of the ellipse

Property 1:

$SG = eSP$, and the tangent and normal at P bisect the external and internal angles between the focal distances of P.

We have $CG = e^2 x'$

Hence $SG = SC + CG = ae + e^2 x' = e \cdot SP$

Also $S'G = CS' - CG = e(a - ex') = eS'P$

Hence $\dfrac{SG}{S'G} = \dfrac{SP}{S'P}$

Therefore, by geometry, PG bisects the angle SPS'

It follows that the tangent bisects the exterior angle between SP and $S'P$

Property 2:

If SY and $S'Y'$ be the $\perp^{lars}$ from the foci upon the tangent at any point P of the ellipse, then Y and Y' lie on the auxiliary circle, and $SY \cdot S'Y' = b^2$. Also CY and $S'P$ are parallel.

The equation to any tangent is

$$x\cos\alpha + y\sin\alpha = P$$

$$(1)$$

Where $P = \sqrt{a^2\cos^2\alpha + b^2\sin^2\alpha}$

The perpendicular SY to (1) passes through the point $(-ae, 0)$ and its equation, is therefore

$$(x + ae)\sin\alpha - y\cos\alpha = 0$$

$$(2)$$

If Y be the point (h, k) then, since Y lies on both (1) and (2), we have

$$h\cos\alpha + k\sin\alpha = \sqrt{a^2\cos^2\alpha + b^2\sin^2\alpha}$$

And $h\sin\alpha - k\cos\alpha = -ae\sin\alpha$

$$= -\sqrt{a^2 - b^2}\sin\alpha$$

Squaring and adding these equations we have $h^2 + k^2 = a^2$, so that Y lies on the auxiliary circle, $x^2 + y^2 = a^2$

Similarly, it may be proved, that Y' lies on this circle

Again S is the point $(-ae, 0)$ and S' is $(ae, 0)$

Hence, from equation (1)

$$SY = P + ae\cos\alpha \text{ and } S'Y' = P - ae\cos\alpha$$

Thus $SY \cdot SY' = P^2 - a^2e^2\cos^2\alpha$

$$= a^2\cos^2\alpha + b^2\sin^2\alpha - (a^2 - b^2)\cos^2\alpha$$

$$= b^2$$

Also $CT = \dfrac{a^2}{CN}$

And therefore, $S'T = \dfrac{a^2}{CN} - ae = \dfrac{a(a - eCN)}{CN}$

$$\frac{CT}{S'T} = \frac{a}{a - e \cdot CN} = \frac{CY}{S'P}$$

Hence, CY and $S'P$ are parallel

Similarly, CY' and SP are parallel

Property 3:

If the normal at any point P meet the major and minor axes in G and g, and if CF be the $\perp^{lar}$ upon this normal then $PF \cdot PG = b^2$ and $PF \cdot Pg = a^2$

The tangent at any point P (the point "ϕ") is

$$\frac{x}{a}\cos\phi + \frac{y}{b}\sin\phi = 1$$

Hence, $PF = \perp$ from C upon this tangent

$$= \frac{1}{\sqrt{\dfrac{\cos^2\phi}{a^2} + \dfrac{\sin^2\phi}{b^2}}} = \frac{ab}{\sqrt{b^2\cos^2\phi + a^2\sin^2\phi}}$$

The normal at P is,

$$\frac{ax}{\cos\phi} - \frac{by}{\sin\phi} = a^2 - b^2$$

If we put $y = 0$, we have $CG = \dfrac{a^2 - b^2}{a}\cos\phi$

$$PG^2 = \left(a\cos\phi - \frac{a^2 - b^2}{a}\cos\phi \right)^2 + b^2\sin^2\phi$$

$$= \frac{b^4}{a^2}\cos^2\phi + b^2\sin^2\phi$$

$$PG = \frac{b}{a}\sqrt{b^2\cos^2\phi + a^2\sin^2\phi}$$

From this and equation (1), we have $PF \cdot PG = b^2$

If we put $x = 0$ in (2), we see that g is the point

$$\left(0, -\frac{a^2 - b^2}{b}\sin\phi\right)$$

Hence $pg^2 = a^2\cos^2\phi + \left(b\sin\phi + \frac{a^2 - b^2}{b}\sin\phi\right)^2$

So that $pg = \frac{a}{b}\sqrt{b^2\cos^2\phi + a^2\sin^2\phi}$

From this result and equation (1), we therefore, have $PF \cdot Pg = a^2$

19. **Let us find the locus of the point of intersection of tangents which meet at right angles.**

Any tangent to the ellipse is

$$y = mx + \sqrt{a^2 m^2 + b^2}$$

And a perpendicular tangent is

$$y = \frac{-1}{m}x + \sqrt{a^2\left(\frac{-1}{m}\right)^2 + b^2}$$

Hence, if (h, k) be their point of intersection, we have

$$k - mh = \sqrt{a^2 m^2 + b^2} \qquad (1)$$

And $mk + h = \sqrt{a^2 + b^2 m^2}$ $\qquad (2)$

If between equations (1) & (2), we eliminate m, we shall have a relation between h and k, squaring and adding these equations, we have

$$(k^2 + h^2)(1 + m^2) = (a^2 + b^2)(1 + m^2)$$

$$b^2 + k^2 = a^2 + b^2$$

Hence, the locus of the point (h, k) is the circle.

$$x^2 + y^2 = a^2 + b^2$$

i.e. a circle whose centre is the centre of the ellipse, and whose radius is the length of the line joining the ends of the major and minor axis. This circle is called the Director circle.

20. Let us prove that through any given point (x_1, y_1) there pass, in general, two tangents to an ellipse.

The equation to any tangent is

$$y = mx + \sqrt{a^2 m^2 + b^2}$$

(1)

If this pass through the fixed point (x_1, y_1), we have

$$y_1 - mx_1 = \sqrt{a^2 m^2 + b^2}$$

Squaring both sides, we get

$$(y_1 - mx_1)^2 = (a^2 m^2 + b^2)$$

$$y_1^2 + m^2 x_1^2 - 2mx_1 y_1 = a^2 m^2 + b^2$$

$$m^2(x_1^2 - a^2) - 2mx_1 y_1 + (y_1^2 - b^2) = 0$$

(2)

For any given values of x_1 and y_1, this equation is in general a quadratic equation, and gives two values of m.

Corresponding to each value of m we have, by substituting in equation (1), a different tangent

The roots of equation (2) are real and different, if

$$(-2x_1 y_1)^2 - 4(x_1^2 - a^2)(y_1^2 - b^2)$$ be positive

i.e. if $b^2 x_1^2 + a^2 y_1^2 - a^2 b^2$ be positive.

$$\frac{x_1^2}{a^2} + \frac{y_1^2}{b^2} - 1$$ be positive.

i.e. if the point (x_1, y_1) be outside the curve.

The roots are equal if

$$b^2 x_1^2 + a^2 y_1^2 - a^2 b^2 \text{ be zero}$$

i.e. if the point (x_1, y_1) lie on the curve.

The roots are imaginary, if

$$\frac{x_1^2}{a^2} + \frac{y_1^2}{b^2} - 1 \text{ be negative,}$$

i.e. if the point (x_1, y_1) lie within the curve.

21. **We will now derive an equation to the chord of contact of tangent drawn from a point** (x_1, y_1)

The equation to the tangent at any point Q, whose co-ordinates are x' and y', is $\dfrac{xx'}{a^2} + \dfrac{yy'}{b^2} = 1$

Also the tangent at the point R, whose co-ordinates are x'' and y'', is

$$\frac{xx''}{a^2} + \frac{yy''}{b^2} = 1$$

If these tangent meet at the point T, whose co-ordinates are x_1 and y_1, we have

$$\frac{x_1 x'}{a^2} + \frac{y_1 y'}{b^2} = 1 \qquad\qquad (1)$$

And $\dfrac{x_1 x''}{a^2} + \dfrac{y_1 y''}{b^2} = 1 \qquad\qquad (2)$

The equation to QR is then

$$\frac{xx_1}{a^2} + \frac{yy_1}{b^2} = 1 \qquad\qquad (3)$$

For, since (1) is true, the point (x', y') lies on (3)

Also, since (2) is true, the point (x'', y''), lies on (3)

Hence, (3) must be the equation to QR the required chord of contact of tangent from (x_1, y_1)

22. To find the equation of the polar of the point (x_1, y_1) with respect to the ellipse $\dfrac{x^2}{a^2} + \dfrac{y^2}{b^2} = 1$

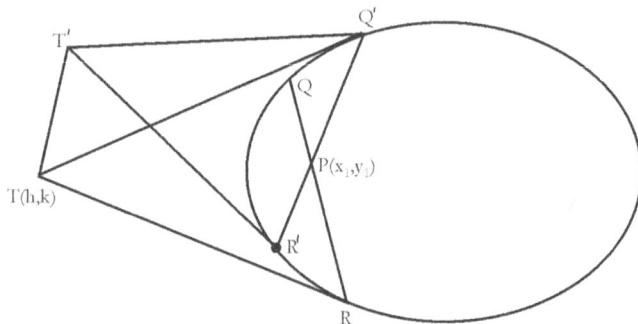

Fig. 107

Let Q and R be the points in which any chord drawn through the point (x_1, y_1) meets the ellipse.

Let the tangents at Q and R meet in the point whose coordinates are (h, k)

We require the locus of (h, k)

Since, QR is the chord of contact of tangents from (h, k) its equation is $\dfrac{xh}{a^2} + \dfrac{yk}{b^2} = 1$

Since this straight line passes through the point (x_1, y_1), we have

$$\frac{hx_1}{a^2} + \frac{ky_1}{b^2} = 1 \qquad\qquad (1)$$

Since, the relation (1) is true, it follows that the point (h, k) lies on the straight line

$$\frac{xx_1}{a^2} + \frac{yy_1}{b^2} = 1 \qquad\qquad (2)$$

Hence, (2) is the equation to the polar of the point (x_1, y_1)

23. **Corollary:** The polar of the focus $(ae, 0)$ is

$$\frac{x \cdot ae}{a^2} = 1$$

$$x = \frac{a^2}{ae}$$

$$x = \frac{a}{e}$$

24. **Observations:**

Point 1:

If the point $P(x_1, y_1)$ lies outside the ellipse the equation to its polar is the same as the equation of the chord of contact of tangent from it.

Point 2:

When (x_1, y_1) is on the ellipse, its polar is the same as the tangent at it.

24. **Let us determine the co-ordinates of the pole of any given line $Ax + By + c = 0$**

Let the line be defined by $Ax + By + c = 0$ (1)

Let (x_1, y_1) be its pole. Then equation (1) must be the same as the polar of (x_1, y_1)

i.e. $\dfrac{xx_1}{a^2} + \dfrac{yy_1}{b^2} - 1 = 0$ (2)

Comparing equations (1) and (2)

The required pole is easily seen to be $\left(\dfrac{-Aa^2}{C}, \dfrac{-Bb^2}{C} \right)$

25. **Let us derive the equation to the pair of tangents that can be drawn to the ellipse from the point (x_1, y_1).**

Let (h, k) be any point on either of the tangents that can be drawn to the ellipse

The equation of the straight line joining (h, k) to (x_1, y_1) is

$$y - y_1 = \frac{k - y_1}{h - x_1}(x - x_1)$$

$$y = y_1 + \frac{(k - y_1)}{h - x_1}(x - x_1)$$

$$y = \frac{y_1(h - x_1) + (k - y_1)(x - x_1)}{h - x_1}$$

$$= \frac{y_1 h - y_1 x_1 + x(k - y_1) - x_1 k + x_1 y_1}{h - x_1}$$

$$y = \frac{k - y_1}{h - x_1}x + \frac{h y_1 - k y_1}{h - x_1}$$

If this straight line touches the ellipse, it must be of the form

$$y = mx + \sqrt{a^2 m^2 + b^2}$$

Hence $m = \dfrac{k - y_1}{h - x_1}$ and $\left(\dfrac{h y_1 - k x_1}{h - x_1}\right)^2 = a^2 m^2 + b^2$

Hence $\left(\dfrac{h y_1 - k x_1}{h - x_1}\right)^2 = a^2 \left(\dfrac{k - y_1}{h - x_1}\right)^2 + b^2$

But this is the condition that the point (h, k) may lie on the locus

$$(xy_1 - x_1 y)^2 = a^2(y - y_1)^2 + b^2(x - x_1)^2 \qquad (1)$$

This equation is therefore the equation to the required tangents.

26. Let us obtain the locus of the middle points of parallel chords of the ellipse.

Let the chords make with the axis an angle whose tangent is m, so that the equation to any one of them, QR is

$$y = mx + c \qquad (1)$$

Where C is different for the different chords

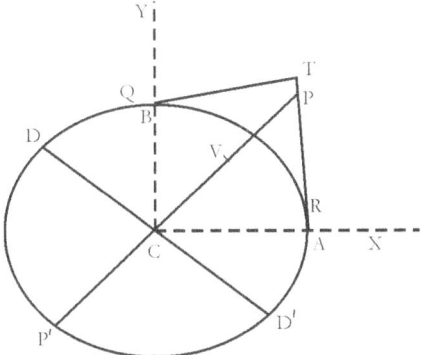

Fig. 108

This straight line meets the ellipse in points whose abscissa are given by the equation

$$\frac{x^2}{a^2} + \frac{y^2}{b^2} = 1$$

Or $\dfrac{x^2}{a^2} + \dfrac{(mx+c)^2}{b^2} = 1$

$$\frac{x^2 b^2 + a^2(m^2 x^2 + c^2 + 2mcx)}{a^2 b^2} = 1$$

Or $x^2 b^2 + a^2 m^2 x^2 + a^2 c^2 + a^2 \cdot 2mcx = a^2 b^2$

Or $x^2(a^2 m^2 + b^2) + 2a^2 mcx + a^2(c^2 - b^2) = 0 \qquad (2)$

Let the roots of this equation

i.e. the abscissa of Q and R, be x_1 and x_2 and let V, the middle point of QR, be the point (h, k)

Then we have

$$h = \frac{x_1 + x_2}{2} = \frac{-a^2 mc}{a^2 m^2 + b^2} \qquad (3)$$

Also v lie on the straight line (1), so that

$$k = mb + c \qquad (4)$$

If between equations (3) and (4), we eliminate C, we have

$$C = k - mb$$

$$\therefore b = \frac{-a^2 m(k - mb)}{a^2 m^2 + b^2}$$

i.e. $b^2 b + a^2 m^2 b = -a^2 mk + a^2 m^2 b$

$$\Rightarrow b^2 b = -a^2 mk \qquad (5)$$

Hence, the point (h, k) always lies on the straight line

$$y = \frac{-b^2}{a^2 m} x \qquad (6)$$

The required locus is therefore the straight line

$$b = m_1 x \text{ where } m_1 = \frac{-b^2}{a^2 m}$$

i.e. $m m_1 = \frac{-b^2}{a^2} \qquad (7)$

27. Let us derive an equation to the chord whose middle point is (b, k).

The required equation is

$$y = mx + c$$

Where m and c are given by

$$c = k - mb \text{ and } m = \frac{-b^2 b}{a^2 k}$$

$$\text{Or } C = \frac{a^2 k^2 + b^2 b^2}{a^2 k}$$

The required equation is therefore

$$y = \frac{-b^2 b}{a^2 k} x + \frac{a^2 k^2 + b^2 b^2}{a^2 k}$$

i.e. $\dfrac{k}{b^2}(y - k) + \dfrac{b}{a^2}(x - b) = 0$

It is therefore $\parallel^{el}$ to the polar of (h, k)

28. **Diameter:**The locus of the middle points of parallel chords of an ellipse is called a diameter, and the chords are called its double ordinates.

29. **Conjugate Diameters:** Two diameters are said to be conjugate when each bisects all chords parallel to the other two diameters.

$y = mx$ and $y = m_1x$ are thereforeConjugate if $mm_1 = \dfrac{-b^2}{a^2}$

30. The tangent at the extremity of any diameter is parallel to the chords which is bisects.

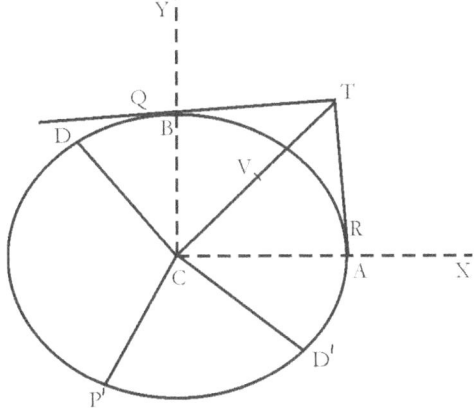

Fig. 109

Let (x', y') be the point P on the ellipse, the tangent at which is parallel to the chord QR, whose equation is

$$y = mx + c \qquad (1)$$

The tangent at the point (x', y') is

$$\frac{xx'}{a^2} + \frac{yy'}{b^2} = 1 \qquad (2)$$

Since equations (1) and (2) are parallel, we have

$$m = \frac{-b^2 x^1}{a^2 y'}$$

i.e. the point (x', y') lies on the straight line

$$y = \frac{-b^2}{a^2 m} x$$

This is the diameter which bisects QR and all chords which are parallel to it.

31. The tangents at the ends of any chord meet on the diameter which bisects the chord.

Let the equation to the chord QR be

$$y = mx + c \tag{1}$$

Let T be the point of intersection of the tangents at Q and R, and let its co-ordinates be h and k.

Since, QR is the chord of contact of tangents from T its equation is

$$\frac{xh}{a^2} + \frac{yk}{b^2} = 1 \tag{2}$$

The equations (1) and (2), therefore represent the same straight line, so that

$$m = \frac{-b^2 h}{a^2 k}$$

i.e. (h, k) lies on the straight line

$$h = \frac{-b^2}{a^2 m} x \text{ which is the equation to the diameter bisecting}$$

the chord QR. Hence T, lies on the straight line CP.

If the eccentric angles of the ends, P and D, of a pair of conjugate diameters be ϕ and ϕ', then ϕ and ϕ' differ by a right angle.

Since, P is the point $(a\cos\phi, b\sin\phi)$, the equation to CP is

$$y = x \cdot \frac{b}{a} \tan\phi \tag{1}$$

So, the equation to CD is

$$y = x \cdot \frac{b}{a} \tan \phi' \qquad (2)$$

These diameters are conjugate if

$$\frac{b^2}{a^2} \tan \phi \tan \phi' = \frac{-b^2}{a^2}$$

i.e. if $\tan \phi = -\cot \phi' = \tan(\phi' \pm 90)$

$$\phi - \phi' = \pm 90°$$

32. **Corollary**: The points on the auxiliary circle corresponding to P and D subtend a right angle at the centre.

For if p and d be these points then, we have

$$\lfloor PCA' = \phi \text{ \& } \lfloor dCA' = \phi'$$

Hence $\lfloor pCd = \lfloor dCA' - \lfloor pCA' = \phi - \phi' = 90°$

If PCP' and DCD' be a pair of conjugate diameters, then (1) $CP^2 + CD^2$ is constant, and (2) The area of the parallelogram formed by the tangents at the ends of these diameters is constant.

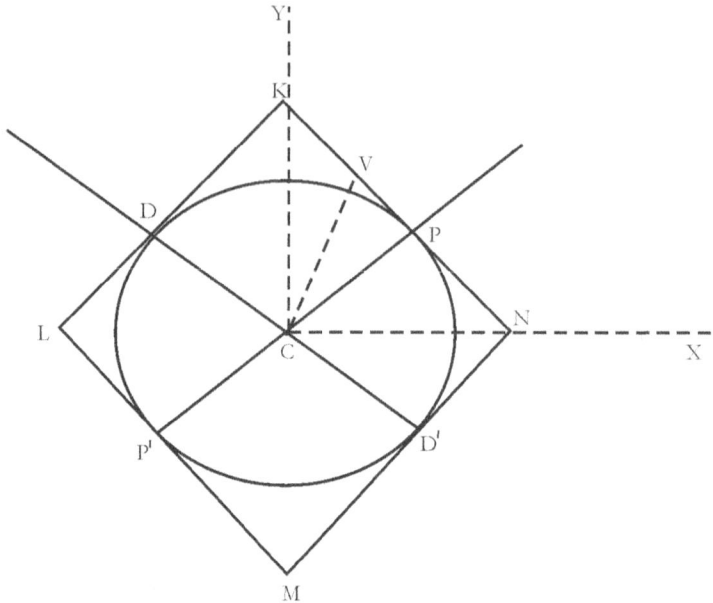

Fig. 110

Let P be the point ϕ, so that its co-ordinates are $a\cos\phi$ and $b\sin\phi$. Then D is the point $90° + \phi$, so that its co-ordinates are $a\cos(90° + \phi)$ and $b\sin(90° + \phi)$ Or $-\sin\phi$ and $b\cos\phi$

(1) We therefore, have

$$CP^2 = a^2\cos^2\phi + b^2\sin^2\phi$$

And $CD^2 = a^2\sin^2\phi + b^2\cos^2\phi$

Hence

$$CP^2 + CD^2 = a^2\cos^2\phi + b^2\sin^2\phi + a^2\sin^2\phi + b^2\cos^2\phi$$

$$= a^2(\cos^2\phi + \sin^2\phi) + b^2(\cos^2\phi + \sin^2\phi)$$

$$= a^2 + b^2$$

= the sum of the squares of the semi-axes of the ellipse.

(2) Let KLMN be the parallelogram formed by the tangents at P, D, P' and D'.

We have are KLMN $= 4 \cdot$ area COKD

$$= 4 \cdot Cu \cdot PK$$

$$= 4 \cdot Cu \cdot CD$$

Where Cu is the $\perp$ from C upon the tangent at P.

Now the equation to the tangent at P is

$$\frac{x}{a}\cos\phi + \frac{y}{b}\sin\phi - 1 = 0$$

So we have

$$Cu = \frac{1}{\sqrt{\dfrac{\cos^2\phi}{a^2} + \dfrac{\sin^2\phi}{b^2}}} = \frac{ab}{\sqrt{a^2\sin^2\phi + b^2\cos^2\phi}} = \frac{ab}{CD}$$

Hence $Cu \cdot CD = ab$

Thus the area of the parallelogram $KLMN = 4ab$

This is equal to the rectangle formed by the tangent at the ends of the major and minor axes

33. The product of the focal distances of a point P is equal to the square on the semi diameter parallel to the tangent at P.

   If P be the point ϕ, then we have

   $$SP = a + ae\cos\phi \text{ and}$$

   $$S'P = (a + ae - \cos\phi)(a - ae\cos\phi)$$

   $$= a^2 - a^2 e^2 \cos^2\phi$$

   $$= a^2 - (a^2 - b^2)\cos^2\phi$$

   $$= a^2 \sin^2\phi + b^2 \cos^2\phi$$

   $$= CD^2$$

34. **Supplemental chords**: The chords joining any point P on an ellipse to the extremities, R and R', of any diameter of the ellipse are called supplemental chords.

35. **Supplemental chords are parallel to conjugate diameters.**

Let P be the point whose eccentric angles is ϕ, and R and R$'$ the points whose eccentric angles are ϕ_1 and $180° + \phi_1$

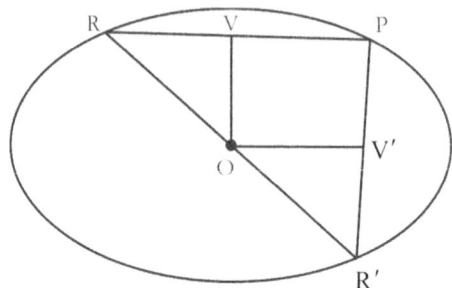

Fig. 111

The equations to PR and PR$'$ and then

$$\frac{x}{a}\cos\left(\frac{\phi+\phi_1}{2}\right)+\frac{y}{b}\sin\left(\frac{\phi+\phi_1}{2}\right)=\cos\left(\frac{\phi-\phi_1}{2}\right) \qquad (1)$$

And $\dfrac{x}{a}\cos\left(\dfrac{\phi+180°+\phi_1}{2}\right)+\dfrac{y}{b}\sin\left(\dfrac{\phi+180°+\phi_1}{2}\right)$

$$=\cos\left(\frac{\phi-180°-\phi_1}{2}\right)$$

i.e. $\dfrac{x}{a}\cos\left(90+\dfrac{\phi+\phi_1}{2}\right)+\dfrac{y}{b}\sin\left(90+\dfrac{\phi+\phi_1}{2}\right)$

$$=\cos\left(\frac{\phi-\phi_1}{2}-90\right)$$

i.e. $\dfrac{-x}{a}\sin\left(\dfrac{\phi+\phi_1}{2}\right)+\dfrac{y}{b}\cos\left(\dfrac{\phi+\phi_1}{2}\right)$

$$=\sin\left(\frac{\phi-\phi_1}{2}\right) \qquad (2)$$

The "m" of the straight line (1) $=\dfrac{-b}{a}\cot\left(\dfrac{\phi+\phi_1}{2}\right)$

The "m" of the line $(2) = \dfrac{b}{a}\tan\left(\dfrac{\phi+\phi_1}{2}\right)$

The product of there m'S $= \dfrac{-b}{a}\cot\left(\dfrac{\phi+\phi_1}{2}\right)\cdot\dfrac{b}{a}\tan\left(\dfrac{\phi+\phi_1}{2}\right)$

$$= \dfrac{-b^2}{a^2}\dfrac{1}{\tan\left(\dfrac{\phi+\phi_1}{2}\right)}\cdot\tan\left(\dfrac{\phi+\phi_1}{2}\right)$$

$$= \dfrac{-b^2}{a^2}$$

The lines PR & PR' are parallel to conjugate diameters.

36. Let us determine the equation to an ellipse referred to a pair of conjugate diameters.

Let the conjugate semi-diameters be CP and CD whose lengths are a' and b' respectively.

If we transform the equation to the ellipse referred to its principal axes, to CP and CD as axes of co-ordinates, then

Since the origin is unaltered, it becomes

$$Ax^2 + 2Hxy + By^2 = 1$$
$$(1)$$

Now the point $P(a',0)$ lies on (1), so that

$$Aa'^2 = 1 \qquad\qquad (2)$$

So since D, the point $(0,b')$, lies on (1), we have

$$Bb'^2 = 1$$

Hence $A = \dfrac{1}{a'^2}$, and $B = \dfrac{1}{b'^2}$

Also, since CP bisects all chords parallel to CD, therefore for each value of x we have two equal and opposite values of y.

This cannot be unless H = 0.

The equation then becomes $\dfrac{x^2}{a'^2} + \dfrac{y^2}{b'^2} = 1$

37. **To prove that, in general, four normal can be drawn from any point to an ellipse, and that the sum of the eccentric angles of their feet is equal to an odd multiple of two right angles.**

The normal at any point, whose eccentric angle is ϕ, is

$$\frac{ax}{\cos\phi} - \frac{by}{\sin\phi} = a^2 - b^2 = a^2 e^2$$

If this normal pass through the point (h, k), we have

$$\frac{ah}{\cos\phi} - \frac{bk}{\sin\phi} = a^2 e^2 \qquad (1)$$

For a given point (h, k) this equation gives the eccentric angles of the feet of the normal which pass through (h, k)

Let $\tan\dfrac{\phi}{2} = t$, so that

$$\cos\phi = \frac{1 - \tan^2\dfrac{\phi}{2}}{1 + \tan^2\dfrac{\phi}{2}} = \frac{1 - t^2}{1 + t^2}$$

And $\sin\phi = \dfrac{2\tan\dfrac{\phi}{2}}{1 + \tan^2\dfrac{\phi}{2}} = \dfrac{2t}{1 + t^2}$

Substituting these values in eqn (1), we have:

$$\frac{ah}{\dfrac{1 - t^2}{1 + t^2}} - \frac{bk}{\dfrac{2t}{1 + t^2}} = a^2 e^2$$

Or $ah\left(\dfrac{1 + t^2}{1 - t^2}\right) - bk\left(\dfrac{1 + t^2}{2t}\right) = a^2 e^2$

i.e. $\dfrac{2tah(1+t^2)-bk(1+t^2)(1-t^2)}{(1-t^2)(2t)}=a^2e^2$

Or $2aht+2aht^3-bk+bkt^4-a^2e^2(2t-2t^3)=0$

Or $bkt^4+2t^3(ah+a^2e^2)+2t(ah-a^2e^2)-bk=0$ $\qquad$ (2)

Let t_1,t_2,t_3 and t_4 be the roots of this equation, so that

$$t_1+t_2+t_3+t_4=-2\dfrac{ah+a^2e^2}{bk} \qquad (3)$$

$$t_1t_2+t_1t_3+t_1t_4+t_2t_3+t_2t_4+t_3t_4=0$$
$$(4)$$

$$t_2t_3t_4+t_3t_4t_1+t_4t_1t_2+t_1t_2t_3=-2\dfrac{ah-a^2e^2}{bk} \qquad (5)$$

And $t_1t_2t_3t_4=-1$ $\qquad$ (6)

Hence, we have

$$\tan\left(\dfrac{\phi_1}{2}+\dfrac{\phi_2}{2}+\dfrac{\phi_3}{2}+\dfrac{\phi_4}{2}\right)=\dfrac{S_1-S_3}{1-S_2+S_4}=\dfrac{S_1-S_2}{0}=\infty$$

$$\therefore \dfrac{\phi_1+\phi_2+\phi_3+\phi_4}{2}n\pi+\dfrac{\pi}{2}$$

And hence $\phi_1+\phi_2+\phi_3+\phi_4=(2n+1)\pi$

$\qquad$ = an odd multiple of two right angle.

Solved Examples

Example 1 : Find the intersection of the tangents at the point ϕ and ϕ'.

The equations to the tangents are

$$\dfrac{x}{a}\cos\phi+\dfrac{y}{b}\sin\phi-1=0$$

And $\dfrac{x}{a}\cos\phi' + \dfrac{y}{b}\sin\phi' - 1 = 0$

Solving these two simultaneous equations

$$
\begin{array}{cccc}
\dfrac{x}{a} & \dfrac{y}{b} & 1 & \\
\sin\phi & +1 & \cos\phi & \sin\phi \\
\sin\phi' & +1 & \cos\phi' & \sin\phi'
\end{array}
$$

$$\therefore \frac{\dfrac{x}{a}}{\sin\phi - \sin\phi'} = \frac{\dfrac{y}{b}}{\cos\phi' - \cos\phi}$$

$$= \frac{-1}{\sin\phi'\cos\phi - \cos\phi'\sin\phi} = \frac{1}{\sin(\phi - \phi')}$$

$$\frac{x}{2a\cos\left(\dfrac{\phi+\phi'}{2}\right)\sin\left(\dfrac{\phi-\phi'}{2}\right)} = \frac{y}{2b\sin\left(\dfrac{\phi+\phi'}{2}\right)\sin\left(\dfrac{\phi-\phi'}{2}\right)}$$

$$= \frac{1}{2\sin\left(\dfrac{\phi-\phi'}{2}\right)\cos\left(\dfrac{\phi-\phi'}{2}\right)}$$

Hence $x = a\,\dfrac{\cos\left(\dfrac{\phi+\phi'}{2}\right)}{\cos\left(\dfrac{\phi-\phi'}{2}\right)}$ & $y = b\,\dfrac{\sin\left(\dfrac{\phi+\phi'}{2}\right)}{\cos\left(\dfrac{\phi-\phi'}{2}\right)}$

12. The Hyperbola

1. **Definition:** The hyperbola is a conic-section in which the eccentricity e is greater than unity.

2. **Let us derive an equation of a hyperbola.**

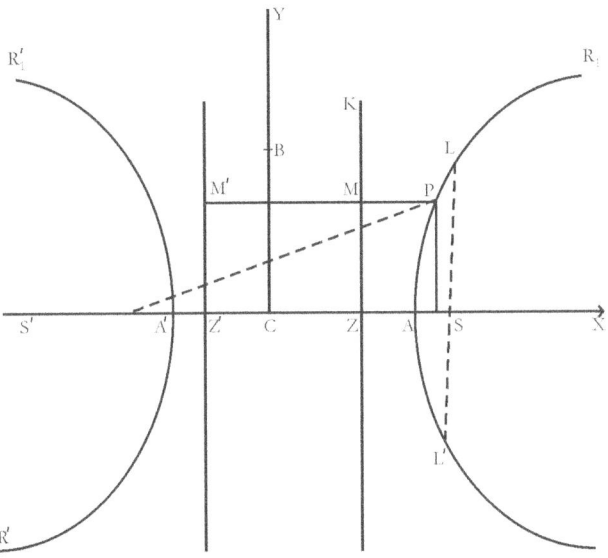

Fig. 112

Let ZK be the directrix, S the focus, and let SZ be perpendicular to the directrix.

There will be a point A on AZ, such that

$$SA = eAZ \tag{1}$$

Since, e > 1, there will be another point A', on SZ produced such that

$$SA' = eA'Z \tag{2}$$

Let the length AA' be called 2a, and let c be the middle point of AA'.

Subtracting (1) from (2), we have

$$SA' - SA = e(A'Z - AZ)$$

$$(\because CA' = CA)$$

$$AA' = 2a$$

$$2a = e \cdot 2CZ$$

$$\therefore CZ = \frac{2a}{2e}$$

Or $CZ = \dfrac{a}{e}$ (3)

Adding (1) & (2), we have

$$SA' + SA = e(A'Z + AZ)$$

$$SA' + SA = e(AA')$$

Or $2ae = (CA' + CS + CS - CA)$

$$2ae = 2CS$$

Or $CS = ae$ (4)

Let C be the origin, CSX the axis of x, and a straight line CY, through $C \perp$ to CX, the axis of y.

Let P be any point on the curve, whose co-ordinates are x and y, and set PM be the $\perp$ upon the directrix, and PN the $\perp$ on AA'

The focus S is the point $(ae, 0)$

The relation $SP^2 = e^2 \cdot PM^2 = e^2 = N^2$ then gives

$$(x - ae)^2 + y^2 = e^2 \left[x - \frac{a}{e} \right]^2$$

i.e. $x^2 - 2aex + a^2e^2 + y^2 = e^2 \left(\dfrac{e^2x^2 + a^2 - 2xae}{e^2} \right)$

i.e. $x^2 - 2aex + a^2e^2 + y^2 = e^2x^2 + a^2 - 2xae$

$$x^2(e^2 - 1) - y^2 = a^2(e^2 - 1)$$

i.e. $\dfrac{x^2}{a^2} - \dfrac{y^2}{a^2(e^2-1)} = 1$ \hfill (5)

$$(\div \text{ by } a^2(e^2-1))$$

Since, in the case of the hyperbola, e > 1 the quantity $a^2(e^2-1)$ is positive. Let it be b^2

So eq (5) becomes

$$\dfrac{x^2}{a^2} - \dfrac{y^2}{b^2} = 1 \hfill (6)$$

Where $b^2 = a^2(e^2-1)$ \hfill (7)

$$= a^2 e^2 - a^2$$

$$= CS^2 - CA^2$$

And therefore $CS^2 = a^2 + b^2$ \hfill (8)

Note:

Eq (6) can be written as

$$\dfrac{y^2}{b^2} = \dfrac{x^2}{a^2} - 1 = \dfrac{x^2 - a^2}{a^2} = \dfrac{(x-a)(x+a)}{a^2}$$

i.e. $\dfrac{PN^2}{b^2} = \dfrac{AN \cdot NA'}{a^2}$

So that $PN^2 : AN \cdot NA' :: b^2 : a^2$

3. **Observations:**

   a. The points A & A' are called the vertices of the hyperbola.

   b. C is the centre

   c. AA' is the transverse axis of the curve the line BB' is called the conjugate axis.

   d. $BC = B'C = b$

   Since S is the point (ae, o), the equation referred to the focus as origin is

---

$$\frac{(x+ae)^2}{a^2} - \frac{y^2}{b^2} = 1$$

$$\frac{x^2 + a^2e^2 + 2aex}{a^2} - \frac{y^2}{b^2} = 1$$

i.e. $\dfrac{x^2}{a^2} + \dfrac{2ex}{a} - \dfrac{y^2}{b^2} + e^2 - 1 = 0$

e. Similarly, the equations, referred to the vertex A and foot of the directrix Z respectively as origins, are found to be

$$\frac{x^2}{a^2} - \frac{y^2}{b^2} + \frac{2x}{a} = 0$$

And $\dfrac{x^2}{a^2} - \dfrac{y^2}{b^2} + \dfrac{2x}{ae} = 1 - \dfrac{1}{e^2}$

There exist a second focus and a second directrix to the curve.

On SC produced take a point S', such that

$$SC = CS' = ae,$$

And another point Z', such that

$$ZC = CZ' = \frac{a}{e}$$

Draw $Z'M' \perp$ to AA', and let PM be produced to meet it in M'

The equation (5) may be written in the form

$$x^2 + 2aex + a^2e^2 + y^2 = e^2x^2 + 2aex + a^2$$

i.e. $(x + ae)^2 + y^2 = e^2\left(\dfrac{a}{e} + x\right)^2$

$$S'P^2 = e^2(Z'C + CN)^2$$

$$S'P^2 = e^2 PM'^2$$

The difference of the focal distances of any point on the hyperbola is equal to the transverse axis.

We have $SP = ePM$ and $S'P = ePM'$

Hence $S'P - SP = e(PM' - PM) = e MM'$

$$= eZZ'$$

$$= 2e CZ$$

$$= 2a$$

$$= \text{the transverse axis } AA'$$

Also $SP = ePM = eZN = e(CN) - e(CZ)$

$$= ex' - a$$

And $S'P = ePM' = e \cdot Z'N$

$$= e \cdot CN + e \cdot Z'C$$

$$= ex' + a$$

Where x' is the abscissa of the point P referred to the center.

4. Latus rectum of the Hyperbola

Let LSL' be the latus rectum, i.e. the double ordinate of the curve drawn through S.

By the definition of the curve, the semi-latus-rectum

SL = e times the distance of L from the dirextrix

$$= eSZ - e(CS - CZ)$$

$$= e CS - e CZ$$

$$= ae^2 - a = \frac{b^2}{a}$$

To trace the curve $\dfrac{x^2}{a^2} - \dfrac{y^2}{b^2} = 1$ (1)

The equation may be written in either of the forms

$$y = \pm b\sqrt{\frac{x^2}{a^2} - 1} \qquad (2)$$

$$x = \pm a \sqrt{\frac{y^2}{b^2} - 1} \qquad (3)$$

From (2), it follow that, if $x^2 < a^2$ i.e. if x lies between a and $-a$ then y is impossible.

Hence there is no curve between A & A'

For all values of $x^2 > a^2$, the equation (2) shows that there are two equal and opposite values of y, so that the curve is symmetrical with respect to the axis of x.

For all values of y, the equation (3) gives two equal and opposite values to x, so that the curve is symmetrical with respect, to the axis of y.

The quantity $\dfrac{x'^2}{a^2} - \dfrac{y'^2}{b^2} - 1$ is positive, zero, or negative according as the point (x', y'), and let the ordinate QN through Q meet the curve in P, so that, by

$$\frac{x^2}{a^2} - \frac{y^2}{b} = 1$$

$$\frac{x'^2}{a^2} - \frac{PN^2}{b^2} = 1$$

And hence $\dfrac{PN^2}{b^2} = \dfrac{x'^2}{a^2} - 1$

If Q be within the curve then y' i.e. QN, is less than PN

So that $\dfrac{y'^2}{b^2} < \dfrac{PN^2}{b^2}$ i.e. $\dfrac{x'^2}{a^2} - 1$

Hence, in this case $\dfrac{x'^2}{a^2} - \dfrac{y'^2}{b^2} > 0$ i.e. positive

Similarly, if Q be without the curve, then $y' > PN$ and

We have $\dfrac{x'^2}{a^2} - \dfrac{y'^2}{b^2} - 1$ negative.

5. **Let us determine the length of any central radius drawn in a given direction.**

We have $\dfrac{x^2}{a^2} - \dfrac{y^2}{b^2} = 1$

Put $x = r\cos\theta$, $y = \sin\theta$, then

$$\dfrac{r^2\cos^2\theta}{a^2} - \dfrac{r^2\sin^2\theta}{b^2} = 1$$

Or $r^2\left(\dfrac{\cos^2\theta}{a^2} - \dfrac{\sin^2\theta}{b^2}\right) = 1$

Or $\dfrac{\cos^2\theta}{a^2} - \dfrac{\sin^2\theta}{b^2} = \dfrac{1}{r^2}$

Or $\dfrac{1}{r^2} = \dfrac{\cos^2\theta}{b^2}\left(\dfrac{b^2}{a^2} - \tan^2\theta\right)$

This equation gives the value of any central radius of the curve drawn at an inclination θ to the transverse axis.

6. **Central conics:**The ellipse and the hyperbola since they both have a center C, such that all chords of the conic passing through its center are bisected at it, are together called central conic.

7. **Polar co-ordinates of hyperbola:**

$$x = a\sec\phi, y = b\tan\phi \text{ satisfies the eq } \dfrac{x^2}{a^2} - \dfrac{y^2}{b^2} = 1$$

Hence there are the suitable polar co-ordinates of hyperbola.

8. **Observations:**The fundamental equation to the hyperbola only differs from that to the ellipse in having $-b^2$ instead of b^2 it will be found that many propositions for the hyperbola are derived from those for the ellipse be changing the sign of b^2.

Point 1: The straight line $y = mx + c$ meet the hyperbola in points which are real, coincident, or imaginary according as $c^2 >=< a^2m^2 - b^2$.

Point 2: The equation to the tangent at (x', y') is
$$\frac{xx'}{a^2} - \frac{yy'}{b^2} = 1$$

Point 3: The straight line $y = mx + \sqrt{a^2 + m^2 - b^2}$ is always a tangent.

Point 4: The straight $x \cos\alpha + y \sin\alpha = P$ is tangent, if
$$p^2 = a^2 \cos^2\alpha - b^2 \sin^2\alpha$$

Point 5: The straight line $lx + my = n$ is a tangent if
$$n^2 = a^2 l^2 - b^2 m^2$$

Point 6: The normal at the point (x', y') is $\dfrac{x - x'}{\dfrac{x'}{a^2}} = \dfrac{y - y'}{\dfrac{y'}{-b^2}}$

9. **Director circle:** The locus of the intersection of tangents which are at right angles is found to be the circle $x^2 + y^2 = a^2 - b^2$, a circle whose centre is the origin and whose radius is $\sqrt{a^2 - b^2}$

    a. If $b^2 < a^2$, this circle is real

    b. If $b^2 = a^2$, the radius of the circle is zero, and it reduces to a point circle at the origin

    c. If $b^2 > a^2$, the radius of the circle is imaginary

10. **Equilateral or Rectangular Hyperbola:** The particular type of hyperbola in which the lengths of the transverse and conjugate axes are equal is called an equilateral or rectangular hyperbola

    Since in this case $b = a$

    The equation to the equilateral hyperbola referred to its centre and axes is $x^2 - y^2 = a^2$

    The eccentricity of the rectangular hyperbola is
    $$e^2 = \frac{a^2 + b^2}{a^2}$$

$$= \frac{a^2 + a^2}{a^2} = \frac{2a^2}{a^2} = 2$$

$$\therefore e = \sqrt{2}$$

11. **Asymptote**: An asymptote is a straight line, which meet the conic in two points both of which are situated at an infinite distance, but which is itself not altogether at infinity.

12. **Let us find the asymptotes of the hyperbola** $\dfrac{x^2}{a^2} - \dfrac{y^2}{b^2} = 1$

The straight line $y = mx + c$ (1)

Meet the hyperbola in points, whose abscissa are given by the equation

$$x^2(b^2 - a^2 m^2) - 2a^2 mcx - a^2(c^2 + b^2) = 0 \quad (2)$$

If the straight line (1) be an asymptote, both roots of (2) must be infinite

Hence, the coefficient of x^2 and x in it must both be zero.

$$\therefore b^2 - a^2 m^2 = 0 \text{ and } a^2 mc = 0$$

$$b^2 = a^2 m^2$$

Or $m^2 = \dfrac{b^2}{a^2}$ and $C = \dfrac{0}{a^2 m}$

$$m = \pm \frac{b}{a} \quad C = 0$$

Substituting these values in (1), we have, as the required equation

$$y = \pm \frac{b}{a} x + 0$$

$$y = \pm \frac{b}{a} x$$

There are therefore two asymptotes both passing through the centre and equally inclined to the axis of x, the inclination being

$$\tan^{-1} \frac{b}{a}$$

The equation to the asymptotes, written as on equation as

$$\frac{x^2}{a^2} - \frac{y^2}{b^2} = 0$$

13. **Corollary:**For all values of C one root of equation (2) is infinite if $m = \pm\frac{b}{a}$. Hence, any straight line, which is parallel to an asymptote meets the curve in one point at infinity and in one finite point.

14. The asymptote passes through two coincident points at infinity touches the cure at infinity, may be seen by finding the equations to the tangents to the curve which pass through any point $\left(x_1, \frac{b}{a}x_1\right)$ on the asymptote $y = \frac{b}{a}x$.

The equation to either tangent through this point is

$$y = mx + \sqrt{a^2 m^2 - b^2}$$

$$\text{Whose } \frac{b}{a}x_1 = mx_1 + \sqrt{a^2 m^2 - b^2}$$

$$\left(\frac{b}{a}x_1 - mx_1\right) = \sqrt{a^2 m^2 - b^2}$$

Squaring both sides, we get

$$\frac{b^2}{a^2}x_1^2 + m^2 x_1^2 - 2\frac{b}{a}mx_1^2 = (a^2 m^2 - b^2)$$

Or $m^2(x_1^2 - a^2) - 2m\frac{b}{a}x_1^2 + \frac{b^2}{a^2}x_1^2 + b^2 = 0$

Or $m^2(x_1^2 - a^2) - 2m\frac{b}{a}x_1^2 + \frac{b^2}{a^2}(x_1^2 + a^2) = 0$

One root of this equation is $m = \frac{b}{a}$, so that, one tangent through the given point is $y = \frac{b}{a}x$

i.e. the asymptote itself.

15. Geometrical construction for the asymptotes:

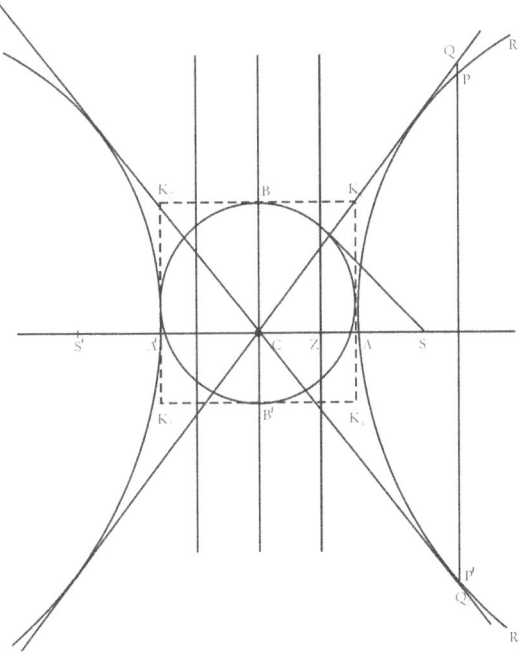

Fig. 113

Let $A'A$ be the transverse axis, and along the conjugate axis measure off CB and CB', each equal to b. Through B and B' draw parallels to the transverse axis and through A and A' parallels to the conjugate axis, and let these meet respectively in k_1, k_2, k_3 and k_4 as in figure.

Clearly the equations of $k_1 ck_3$ & k_2 are

$$y = \frac{b}{a}x \,\&\, y = \frac{-b}{a}x$$

There are called equations of the asymptotes.

Let any double ordinate PNP' of the hyperbola be produced both ways to meet the asymptotes in Q and Q', and let the abscissa CN be x'.

Since P lies on the curve, we have

$$NP = \frac{b}{a}\sqrt{x'^2 - a^2}$$

Since Q is on the asymptote whose equation is $y = \frac{b}{a}x$

We have $NQ = \frac{b}{a}x' \; (\because NQ = y)$

Hence $PQ = NQ - NP$

$$= \frac{b}{a}x' - \frac{b}{a}\sqrt{x'^2 - a^2}$$

$$= \frac{b}{a}\left(x'\sqrt{x'^2 - a^2}\right)$$

And $P'Q = NP' + NQ$

$$= \frac{b}{a}\sqrt{x'^2 - a^2} + \frac{b}{a}x' = \frac{b}{a}\left(x' + \sqrt{x'^2 - a^2}\right)$$

Therefore $PQ \cdot P'Q = \frac{b}{a}\left(x' - \sqrt{x'^2 - a^2}\right) \cdot \frac{b}{a}\left(x' + \sqrt{x'^2 - a^2}\right)$

$$= \frac{b^2}{a^2}\left\{(x')^2 - \left(\sqrt{x'^2 - a^2}\right)^2\right\}$$

$$= \frac{b^2}{a^2}\left\{x'^2 - x'^2 + a^2\right\}$$

$$= \frac{b^2}{a^2} \times a^2$$

$$= b^2$$

Again, $PQ = \frac{b}{a}\left(x' - \sqrt{x'^2 - a^2}\right)$

$$= \frac{b}{a}\left(x' - \sqrt{x'^2 - a^2}\right) \times \frac{\left(x' + \sqrt{x'^2 - a^2}\right)}{\left(x' + \sqrt{x'^2 - a^2}\right)}$$

$$= \frac{b}{a} \frac{\left[x'^2 - (x'^2 - a^2)\right]}{\left(x' + \sqrt{x'^2 - a^2}\right)}$$

$$= \frac{b}{a} \times \frac{a^2}{x' + \sqrt{x'^2 - a^2}}$$

$$= \frac{ab}{x' + \sqrt{x'^2 - a^2}}$$

PQ is always positive.

If x' is infinitely great, PQ is infinitely small.

If SF be the $\perp^{lar}$ from S upon on asymptote, the point F lies on the auxiliary circle. This follows from the fact that the asymptote is a tangent, whose point of contact happens to lie at infinity, or it may be proved directly.

For $CF = CS \cos FCS$

$$= CS \frac{CA}{CK}$$

$$= \sqrt{a^2 + b^2} \cdot \frac{a}{\sqrt{b^2 + b^2}} = a$$

Also,

Z being the foot of the directrix, we have

$$CA^2 = CS \cdot CZ$$

And hence $CF^2 = CS \cdot CZ$

i.e. $CS : CF :: CF : CZ$

By geometry, it follows that $\lfloor CZF = \lfloor CFS = a$ right angle and hence that F lies on the directrix

---

Hence, the perpendiculars from the foci on either asymptote meet it in the same points as the corresponding directrix, and the common points of intersection lie on the auxiliary circle.

15. **Equilateral or Rectangular Hyperbola**: In this curve the quantities a and b are equal. The equations to the asymptotes are therefore $y = \pm x$.

16. **Conjugate Hyperbola:**

The hyperbola $\dfrac{y^2}{b^2} - \dfrac{x^2}{a^2} = 1$ is conjugate to the hyperbola $\dfrac{x^2}{a^2} - \dfrac{y^2}{b^2} = 1$

The equations to the asymptotes of $\dfrac{y^2}{b^2} - \dfrac{x^2}{a^2} = 1$ and

$$\dfrac{x^2}{a^2} - \dfrac{y^2}{b^2} = 1 \text{ is } \dfrac{y^2}{b^2} - \dfrac{x^2}{a^2} = 0$$

Thus, a hyperbola and it conjugate have the same asymptotes.

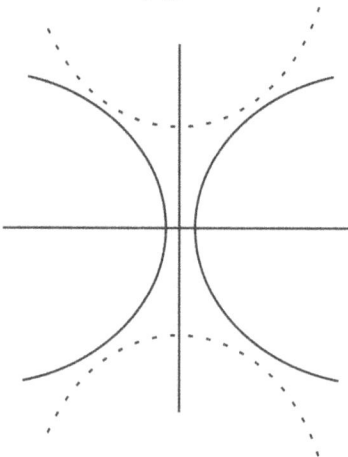

Fig. 114

The conjugate hyperbola is the dotted curve in the figure above.

17. Let us determine the intersection of a hyperbola with a pair of conjugate diameters.

The straight line $y = m_1 x$ intersect the hyperbola

$$\frac{x^2}{a^2} - \frac{y^2}{b^2} = 1$$

In points whose abscissas are given by

$$x^2 \left[\frac{1}{a^2} - \frac{m_1^2}{b^2} \right] = 1$$

Or $x^2 \left(\frac{b^2 - a^2 m_1^2}{a^2 b^2} \right) = 1$

i.e. $x^2 = \dfrac{a^2 b^2}{a^2 - a^2 m_1^2}$

The points are therefore real or imaginary, according as

$$a^2 m_1^2 \text{ is} < \text{ or } > b^2$$

i.e. according as

$$m_1 \text{ is numerically } < \text{ or } > \frac{b}{a} \qquad (1)$$

The straight lines $y = m_1 x$ and $y = m_2 x$ are conjugate diameters if

$$m_1 m = \frac{b^2}{a^2} \qquad (2)$$

Hence, one of the quantities m_1 & m_2 must be less than $\dfrac{b}{a}$ and the other greater than $\dfrac{b}{a}$

Let m_1 be $< \dfrac{b}{a}$, so that, by (1), the straight line $y = m_1 x$ meets the hyperbola in real points

Then by (2), m_2 must be $> \dfrac{b}{a}$, so that, by (1), the straight line $y = m_2x$ will meet the hyperbola in imaginary points

Only one pair of conjugate diameters meets a hyperbola in real points.

18. **If a pair of diameters be conjugate with respect to a hyperbola, they will be conjugate with respect to its conjugate hyperbola.**

For the straight line $y = m_1x$ and $y = m_2x$ are conjugate with respect to the hyperbola

$$\frac{x^2}{a^2} - \frac{y^2}{b^2} = 1 \qquad (1)$$

If $m_1 m_2 = \dfrac{b^2}{a^2}$ $\qquad (2)$

The conjugate hyperbola is $\dfrac{y^2}{b^2} - \dfrac{x^2}{a^2} = 1$

Which is obtained by replacing $a^2 = -a^2$ and $b^2 = -b^2$ in (1)

Hence $m_1 m_2 = \dfrac{-b^2}{-a^2}$

$\qquad = \dfrac{b^2}{a^2} \qquad (3)$

The relation (3) is the same as (2)

19. **If a pair of diameters be conjugate with respect to a hyperbola, one of them meets the hyperbola in real points and the other meets the conjugate hyperbola in real points.**

Let the diameters by $y = m_1x$ and $y = m_2x$, so that

$$m_1 m_2 = \frac{b^2}{a^2}$$

Let $m_1 < \dfrac{b}{a}$ and hence, $m_2 > \dfrac{b}{a}$, so that the straight line $y = m_1 x$
meets the hyperbola in real points.

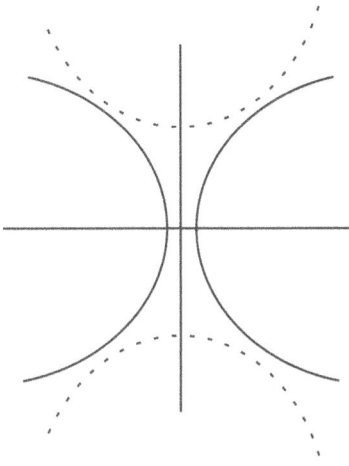

Fig. 115

Also, the straight line $y = m_2 x$ meet the conjugate hyperbola
$\dfrac{y^2}{b^2} - \dfrac{x^2}{a^2} = 1$ in points whose abscissas are given by the equation

$$x^2 \left(\dfrac{m_2^2}{b^2} - \dfrac{1}{a^2} \right) = 1$$

i.e. $x^2 \left(\dfrac{a^2 m_2^2 - b^2}{b^2 a^2} \right) = 1$

Or $x^2 = \dfrac{b^2 a^2}{a^2 m_2^2 - b^2}$

Since $m_2 > \dfrac{b}{a}$, there abscissas are real.

20. **If a pair of conjugate diameters meet the hyperbola and its conjugate in P and D, then (1) $CP^2 - CD^2 = a^2 - b^2$, and the**

tangents at **P, D** and the other ends of the diameters passing through them form a parallelogram whose vertices lie on the asymptotes and whose area is constant.

Let P be any point on the hyperbola $\dfrac{x^2}{a^2} - \dfrac{y^2}{b^2} = 1$ whose co-ordinates are $(a\sec\phi, b\tan\phi)$

The equation to the diameter CP is therefore

$$y = \frac{b\tan\phi}{a\sec\phi} \cdot x$$

$$= \frac{b\dfrac{\sin\phi}{\cos\phi}}{a \cdot \dfrac{1}{\cos\phi}} \cdot x$$

$$= \frac{b\sin\phi \cdot x}{a}$$

$$\therefore y = x \cdot \frac{b}{a}\sin\phi$$

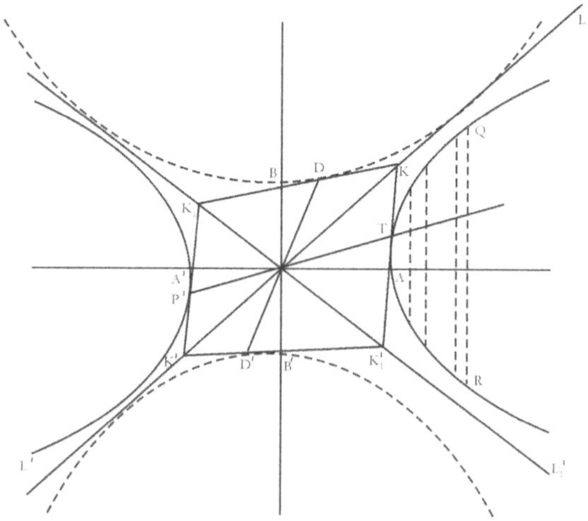

Fig. 116

The equation to the straight line, which is conjugate to CP is

$$y = x \frac{b}{a \sin \phi}$$

This straight line meets the conjugate hyperbola $\dfrac{y^2}{b^2} - \dfrac{x^2}{a^2} = 1$ in the points $(a \tan \phi, b \sec \phi)$ and $(-a \tan \phi, -b \sec \phi)$ so that D is the point $(a \tan \phi, b \sec \phi)$

We therefore, have

$$CP^2 = a^2 \sec^2 \phi + b^2 \tan^2 \phi$$

And $CD^2 = a^2 \tan^2 \phi + b^2 \sec^2 \phi$

Hence,

$$CP^2 - CD^2 = (a^2 \sec^2 \phi - a^2 \tan^2 \phi) + (b^2 \tan^2 \phi - b^2 \sec^2 \phi)$$
$$= a^2 (\sec^2 \phi - \tan^2 \phi) + b^2 (\tan^2 \phi - \sec^2 \phi)$$
$$= \sec^2 \phi - \tan^2 \phi (a^2 - b^2)$$
$$= a^2 - b^2 \; (\because \sec^2 \phi - \tan^2 \phi = 1)$$

Again, the tangents at P and D to the hyperbola and the conjugate hyperbola are easily seen to be

$$\frac{x}{a} - \frac{y}{b} \sin \phi = \cos \phi \qquad (1)$$

$$\frac{y}{b} - \frac{x}{a} \sin \phi = \cos \phi \qquad (2)$$

These meet at the point

$$\begin{array}{ccc} \dfrac{x}{a} & \dfrac{y}{b} & 1 \end{array}$$

$$\begin{array}{cccc} -\sin \phi & -\cos \phi & 1 & -\sin \phi \\ 1 & -\cos \phi & -\sin \phi & 1 \end{array}$$

$$\frac{\dfrac{x}{a}}{\sin \phi \cos \phi + \cos \phi} = \frac{\dfrac{y}{b}}{\sin \phi \cos \phi + \cos \phi} = \frac{1}{1 - \sin^2 \phi}$$

Or $\dfrac{x}{a} = \dfrac{(\sin\phi+1)\cos\phi}{(1+\sin\phi)(1-\sin\phi)}$, $\dfrac{y}{b} = \dfrac{(\sin\phi+1)\cos\phi}{(1+\sin\phi)(1-\sin\phi)}$

$\therefore \dfrac{x}{a} = \dfrac{y}{b} = \dfrac{\cos\phi}{1-\sin\phi}$

This point lies on the asymptote CL.

Similarly, the intersection of the tangents at P and D' lies on CL_1', that of tangents at D' & P' on CL', and those at D and P' on CL_1

The perpendicular from C on the straight line (1)

$= \dfrac{\cos\phi}{\sqrt{\dfrac{1}{a^2}+\dfrac{1}{b^2}\sin^2\phi}} = \dfrac{\cos\phi}{\dfrac{\sqrt{b^2+a^2\sin^2\phi}}{a^2b^2}}$

$= \dfrac{ab\cos\phi}{\sqrt{b^2+a^2\sin^2\phi}}$

$= \dfrac{1}{\cos\phi}\dfrac{ab}{\sqrt{a^2+a^2\sin^2\phi}}$

$= \dfrac{ab}{\sqrt{\dfrac{b^2}{\cos^2\phi}+\dfrac{a^2\sin^2\phi}{\cos^2\phi}}} = \dfrac{ab}{\sqrt{b^2\sec^2\phi+a^2\tan^2\phi}}$

$= \dfrac{ab}{CD} = \dfrac{pb}{PK}$

So that PKX perpendicular from C on $PK = ab$

i.e. area of the parallelogram $CPKD = ab$

Also, the areas of the parallelograms CPKD, CDK_1P', $CP'K'D'$, and $CD'K_1'P$ are all equal.

The area $KK_1K'K_1'$ therefore $= 4ab$

21. **Corollary:** $PK = CD = D'C = K_1'P$, so that the portion of a tangent to a hyperbola intercepted between the asymptotes is bisected at the point of contact.

22. Whatever be the form of the equation to a hyperbola, the equation to the asymptotes only differs from it by a constant, and the equation to the conjugate hyperbola differs from that to the asymptotes by the same constant.

23. **Let us find the equation to a hyperbola referred to its asymptotes.**

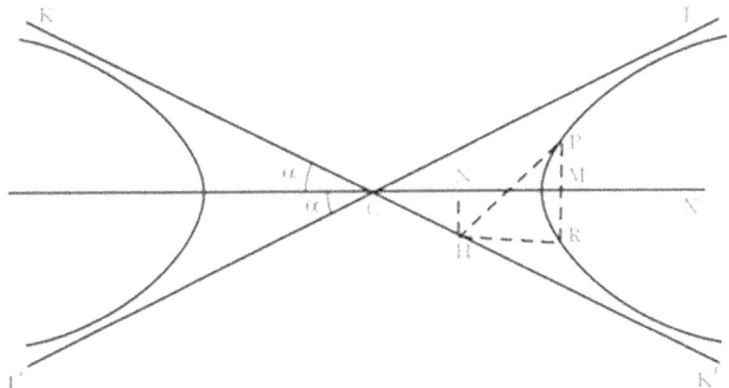

Fig. 117

Let P be any point on the hyperbola, whose equation referred to its axes is $\dfrac{x^2}{a^2} - \dfrac{y^2}{b^2} = 1$ (1)

Draw PH $\parallel^{el}$ to one asymptote CL to meet the other CK' IN H, and let CH and HP be h and k respectively.

Then h and k are the co-ordinates of P referred to the asymptotes.

Let α be the semi-angle between the asymptotes so that

$\tan \alpha = \dfrac{b}{a}$

And hence $\dfrac{\sin\alpha}{\cos\alpha}=\dfrac{b}{a}$

Or $\dfrac{\sin\alpha}{b}=\dfrac{\cos\alpha}{a}=\dfrac{1}{\sqrt{a^2+b^2}}$

Draw HN $\perp^{lar}$ to the transverse axis, and HR parallel to the transverse axis, to meet the ordinate PM of the point P in R.

Then, since PH & HR are parallel respectively to CL and CM, we have

$$\underline{|PHR}=\underline{|LCM}=\alpha$$

Hence $CM = CN + HR$

$$= CH\cos\alpha + HP\cos\alpha$$
$$= \cos\alpha(CH + HP)$$
$$= \cos\alpha(CH + HP)$$
$$= \cos\alpha(h + k)\left(\because \cos\alpha = \dfrac{a}{\sqrt{a^2+b^2}}\right)$$
$$= \dfrac{a}{\sqrt{a^2+b^2}}(h + k)$$

And $MP = RP - HN$

$$= HP\sin\alpha - CH\sin\alpha$$
$$= (k - h)\dfrac{b}{\sqrt{a^2+b^2}}$$

Therefore, since CM and MP satisfy the equation (1), we have

$$\dfrac{x^2}{a^2}-\dfrac{y^2}{b^2}=1$$

$$\dfrac{\left(\dfrac{a(h+k)}{\sqrt{a^2+b^2}}\right)^2}{a^2}-\dfrac{\left(\dfrac{b(k-h)}{\sqrt{a^2+b^2}}\right)^2}{b^2}=1$$

$$\frac{(h+k)^2}{a^2+b^2} - \frac{(k-h)^2}{a^2+b^2} = 1$$

$$(h+k)^2 - (k-h)^2 = a^2 + b^2$$

$$h^2 + k^2 + 2hk - k^2 - h^2 + 2hk = a^2 + b^2$$

Or $4hk = a^2 + b^2 \Rightarrow hk = \dfrac{a^2+b^2}{4}$

Hence, since (h,k) is any point on the hyperbola, the required equation is

$$xy = \frac{a^2+b^2}{4}$$

Similarly, the equation to the conjugate hyperbola, referred to the asymptotes

$$xy = -\frac{a^2+b^2}{4}$$

24. **Let us find the equation to the tangent at any point of the hyperbola** $xy = c^2$

Let (x', y') be any point p on the hyperbola, and (x'', y'') a point Q on it, so that we have

$$x'y' = c^2 \tag{1}$$

$$\Rightarrow y' = \frac{c^2}{x'}$$

And $x''y'' = c^2$ $\tag{2}$

$$\Rightarrow y'' = \frac{c^2}{x''}$$

The equation to the line PQ is then

$$y - y' = \frac{y'' - y'}{x'' - x'}(x - x') \tag{3}$$

But, by (1) & (2), we have

$$\frac{y'' - y'}{x'' - x'} = \frac{\dfrac{c^2}{x''} - \dfrac{c^2}{x'}}{x'' - x'}$$

$$= \frac{c^2 \left(\dfrac{x' - x''}{x'' x'} \right)}{x'' - x'}$$

$$= \frac{c^2}{x'' x'} \left[\frac{-(x'' - x')}{x'' - x'} \right] = \frac{-c^2}{x'' x'}$$

Hence, the equation (3) becomes

$$y - y' = \frac{-c^2}{x'' x'} (x - x') \qquad (4)$$

Let now the point Q be taken indefinitely near to P, so that $x'' = x'$ ultimately, and therefore PQ becomes the tangent at P. Then (4), becomes

$$y - y' = \frac{-c^2}{(x')^2} (x - x')$$

$$= \frac{-y'}{x'} (x - x') \; (\because \text{by (1)} \; c^2 = x'y')$$

The required equation is therefore

$$xy' + x'y = 2x'y' = 2c^2 \qquad (5)$$

The equation (5) may also e written in the form

$$\frac{x}{x'} + \frac{y}{y'} = 2 \; (\div \text{ by } x'y') \qquad (6)$$

25. The tangent at any point of a hyperbola cuts off a triangle of constant area from the asymptotes, and the portion of it intercepted between the asymptotes is bisected at the point of contact.

Take the asymptotes as axes and let the equation to the hyperbola be $xy = c^2$

The tangent at any point P is $\dfrac{x}{x'} + \dfrac{y}{y'} = 2$

This meet the axes in the points $(2x', 0)$ & $(0, 2y')$

If there points be L and L', and the centre be C, we have

$$CL = 2x' \text{ and } CL' = 2y'$$

If 2α be the angle between the asymptotes, the area of the triangle

$$LCL' = \frac{1}{2} CL \cdot CL' \sin 2\alpha$$

$$= 2x'y' \sin 2\alpha$$

$$= \frac{a^2 + b^2}{2} \cdot 2 \sin \alpha \cos \alpha$$

$$= \frac{a^2 + b^2}{2} \times 2 \times \frac{b}{\sqrt{a^2 + b^2}} \times \frac{a}{\sqrt{a^2 + b^2}}$$

$$= ab$$

Also since L is the point $(2x', 0)$ & L' is $(0, 2y')$ the middle point of LL' is (x', y'). i.e. the point of contact p

The polar of any point (x_1, y_1) with respect to the curve is

$$xy_1 + x_1 y = 2c^2$$

Since the point (x_1, y_1) does not lie on the curve it cannot be written in the form $\dfrac{x}{x'} + \dfrac{y}{y'} = 2$

26. **The equation to the normal at the point (x', y') is**
    $y - y' = m(x - x')$**, where m is chosen so that this line is $\perp$ to the tangent**

$$y = \frac{-y'}{x'} x + \frac{2c^2}{x'}$$

If w be the angle between the asymptotes we then obtain,

$$m = \frac{x' - y' \cos w}{y' - x' \cos w}$$

So that the required equation to the normal is

$$y(y' - x' \cos w) - x(x' - y' \cos w) = y'^2 - x'^2$$

Also $\cos w = \cos 2\alpha = \cos^2 \alpha - \sin^2 \alpha$

$$= \frac{a^2}{a^2 + b^2} - \frac{b^2}{a^2 + b^2}$$

$$= \frac{a^2 - b^2}{a^2 + b^2}$$

If the hyperbola be rectangular, then $w = 90°$

$$\cos 90 = 0$$

The equation to the normal becomes

$$y(y' - x'(0)) - x(x' - y'(0)) = y'^2 - x'^2$$

Or $xx' - yy' = x'^2 - y'^2$

27. Let us obtain an equation of the asymptotes referred to one variable.

The equation $xy = c^2$ is clearly satisfied by the substitution $x = ct$ and $y = \frac{c}{t}$.

Hence, for all values of t, the point whose co-ordinates are $\left(ct, \frac{c}{t}\right)$ lies on the curve, and it may be called the point "t".

The tangent at the point "t" is $\frac{x}{t} + yt = 2c$

Also the normal is,

$$y(1 - t^2 \cos w) - x(t^2 - \cos w) = \frac{c}{t}(1 - t^4)$$

Or when the hyperbola is rectangular

$$y - xt^2 = \frac{c}{t}(1-t^4)\ (\because \cos w = 0)$$

The equations to the tangent at the point " t_1 " and " t_2 " are

$$\frac{x}{t_1} + yt_1 = 2c \text{ and } \frac{x}{t_2} + yt_2 = 2c$$

And hence the tangents meet at the point

$$\left(\frac{2ct_1t_2}{t_1 + t_2}, \frac{2c}{t_1 + t_2} \right)$$

The line joining " t_1 " and " t_2 " which is the polar of this point is therefore

$$x + yt_1t_2 = C(t_1 + t_2)$$

The equations to the line joining the points are

$$\left(ct_1, \frac{c}{t_1} \right) \ \& \ \left(ct_2, \frac{c}{t_2} \right)$$

Solved Problems

Example 1: The perpendiculars from the centre upon the tangent and normal at any point of the hyperbola $\dfrac{x^2}{a^2} - \dfrac{y^2}{b^2} = 1$ **meet them in Q and R. find the loci of Q and R.**

The straight line $x \cos \alpha + y \sin \alpha = p$ is a tangent if

$$p^2 = a^2 \cos^2 \alpha - b^2 \sin^2 \alpha$$

But p and α are the polar co-ordinates of Q, the foot of the perpendicular on this straight line from C.

The polar equation to the locus of Q is therefore

$$r^2 = a^2 \cos\theta - b^2 \sin^2\theta\ (\because p = r \ \& \ \alpha = \theta)$$

In Cartesian co-ordinates

$$(x^2 + y^2)^2 = a^2 x^2 - b^2 y^2$$

If the hyperbola be rectangular, we have $a = b$, and the polar equation is

$$r^2 = a^2 \cos^2 \theta - a^2 \sin^2 \theta \; (\because b = a)$$
$$= a^2 (\cos^2 \theta - \sin^2 \theta)$$
$$= a^2 \cos 2\theta$$

Again any normal is

$$ax \sin \phi + by = (a^2 + b^2) \tan \phi$$
$$(1)$$

The equation to the perpendicular on it from the origin is

$$bx - ay \sin \phi = 0$$
$$(2)$$

If we eliminate ϕ, we have the locus of R from (2), we have

$$\sin \phi = \frac{bx}{ay}$$

And then $\tan \phi = \dfrac{\sin \phi}{\cos \phi} = \dfrac{\sin \phi}{\sqrt{1 - \sin^2 \phi}}$

$$= \frac{\dfrac{bx}{ay}}{\sqrt{1 - \dfrac{b^2 x^2}{a^2 y^2}}} = \frac{\dfrac{bx}{ay}}{\dfrac{\sqrt{a^2 y^2 - b^2 x^2}}{ay}}$$

$$= \frac{bx}{\sqrt{a^2 y^2 - b^2 x^2}}$$

Substituting in (1), the locus is

$$ax \times \frac{bx}{ay} + by = (a^2 + b^2) \frac{bx}{\sqrt{a^2 y^2 - b^2 x^2}}$$

$$b\left(\frac{x^2 + y^2}{y} \right) = \frac{(a^2 + b^2) \cdot bx}{\sqrt{a^2 y^2 - b^2 x^2}}$$

Or $(x^2 + y^2)\sqrt{a^2 y^2 - b^2 x^2} = (a^2 + b^2)x\,y$

Squaring both sides, we get

$$(x^2 + y^2)^2(a^2 y^2 - b^2 x^2) = (a^2 + b^2)^2 x^2 y^2$$

Example 2: Find the asymptotes of the hyperbola $3x^2 - 5xy - 2y^2$
$+5x + 11y - 8 = 0$

The equation of the hyperbola is $3x^2 - 5xy - 2y^2 + 5x + 11y - 8 = 0$

$\qquad$ (1)

Since the equation to the asymptotes only differs from it by a constant, it must be of the from

$$3x^2 - 5xy - 2y^2 + 5x + 11y + c = 0$$

$\qquad$ (2)

Since, (2) represents the asymptotes it must represent two straight lines. We know that the general equation of the second degree

$$ax^2 + 2hxy + by^2 + 2gx + 2fy + c = 0$$

May represent two straight lines if $abc + 2fgh - af^2 - bg^2 - ch^2 = 0$

From (1) & (2)

$$3(-2)c + 2 \cdot \frac{5}{2} \cdot \frac{11}{2}\left(\frac{-5}{3}\right) - 3\left(\frac{11}{2}\right)^2 - (-2)\left(\frac{5}{2}\right)^2 - c\left(\frac{-5}{2}\right)^2 = 0$$

$$-6c - \frac{55 \times 5}{6} - \frac{3 \times 121}{4} + \frac{25}{2} - \frac{25c}{4} = 0$$

$$\therefore C = -12$$

The equation to the asymptotes is therefore,

$$3x^2 - 5xy - 2y^2 + 5x + 11y - 12 = 0$$

And the equation to the conjugate hyperbola is

$$3x^2 - 5xy - 2y^2 + 5x + 11y - 16 = 0$$

The equation to any hyperbola whose asymptotes are the straight lines

$$Ax + By + C = 0\ \&\ A_1 x + B_1 y + C_1 = 0 \text{ is}$$

$$(Ax + By + C)(A_1x + B_1y + C_1) = \lambda^2 \tag{3}$$

Where λ is any constant:

For (3), only differs by a constant from the equation to the asymptotes, which is

$$(Ax + By + C)(A_1x + B_1y + C_1) = 0 \tag{4}$$

If in (3), we substitute $-\lambda^2$ for λ^2 we shall have the equation to the conjugate hyperbola

It follows that any equation of the form

$$(Ax + By + C)(A_1x + B_1y + C_1) = \lambda^2$$

Represents a hyperbola whose asymptotes are

$$Ax + By + C = 0 \text{ and } A_1x + B_1y + C_1 = 0$$

Thus the equation $x(x + a) = a^2$ represents a hyperbola whose asymptotes are $x = 0$ and $x + y = 0$

Again, the equation $x^2 + 2xy \cot 2\alpha - y^2 = a^2$

i.e. $(x \cot \alpha - y)(x \tan \alpha + y) = a^2$

Represents a hyperbola whose asymptotes are

$$(x \cot \alpha - y) = 0 \text{ and } x \tan \alpha + y = 0$$

13.   Curve Tracing

1.   Polar equation: Relation between the polar co-ordinates of any point on the curve

Let ZS be chosen as the positive direction of the initial line, and produce it to X.

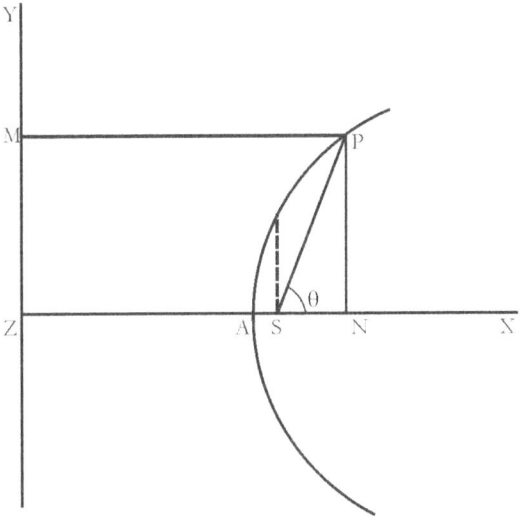

Fig. 118

Take any point P on the curve, and let its polar co-ordinates be r and θ, so that we have

$$SP = r \text{ and } \underline{|XSP} = \theta$$

Draw PN $\perp$ to the initial line, and PM $\perp$ to the directrix.

Let SL be the semi-latus-rectum, and let $SL = l$

Since $SL = e \cdot SZ$, we have

$$SZ = \frac{l}{e}$$

Hence $r = SP = ePM = eZN$

1. $= e(ZS + SN)$

$$= e\left(\frac{l}{e} + SP\cos\theta\right)\left(\because \frac{SN}{SP} = \cos\theta\right)$$

$r = l + e \cdot r\cos\theta$

Therefore $r(1 - e\cos\theta) = l$

Or $r = \dfrac{l}{1 - e\cos\theta}$ \hfill (1)

2. **Corollary:** If SZ be taken as the positive direction of the initial line and the vectorial angle measured clockwise, the equation to the curve is

$$r = \frac{l}{1 + e\cos\theta}$$

If the conic be a parabola then $e = 1$, And the equation is

$$r = \frac{l}{1 - e\cos\theta}$$

$$= \frac{l}{1 - \cos\theta}$$

$$= \frac{l}{2\sin^2\dfrac{\theta}{2}}$$

$$= \frac{l}{2}\csc^2\frac{\theta}{2}$$

If the critical line, instead of being the axis, be such that the axis is inclined at an angle γ to it, then instead of θ we must substitute $\theta - \gamma$

The equation in this case is then $\dfrac{l}{\gamma} = 1 - e\cos(\theta - \gamma)$

3. **To trace the curve** $\dfrac{l}{\gamma} = 1 - e\cos\theta$

Case 1: If $e = 1$, then equation becomes $\dfrac{l}{r} = 1 - \cos\theta$

θ	$\cos\theta$	$\dfrac{l}{\gamma}$	γ
If $\theta = 0°$	1	0	γ is infinite
If $\theta = 0°$ to $90°$	Decrease from 1 to 0	Increases from 0 to 1	γ decreases form infinity to 1.
If $\theta = 90°$ to $180°$	Decrease from 0 to -1	Increase from 1 to 2	Decreases from l to $\dfrac{l}{2}$
If $\theta = 180°$ to $270°$	Increase from -1 to 0	Decrease from 2 to 1	Increases $\dfrac{l}{2}$ to l
If $\theta = 270°$ to $360°$	Increase from 0 to 1	Decrease from 1 to 0	γ increase from l to ∞

The curve is thus the parabola $\infty\,\text{FPLA}\;L'P'F'\;\infty$

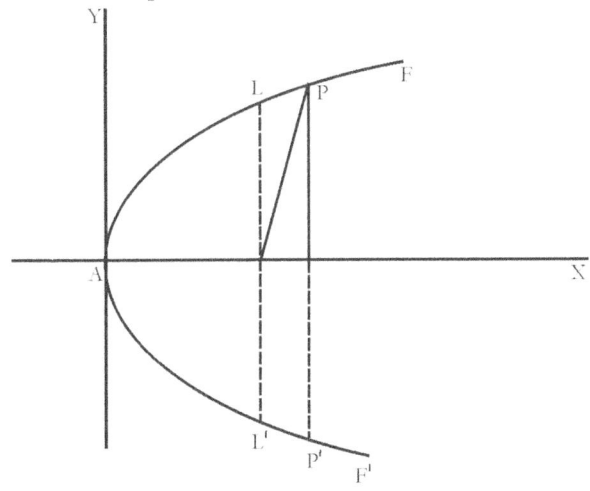

Fig. 119

Case 2: If $e < 1$

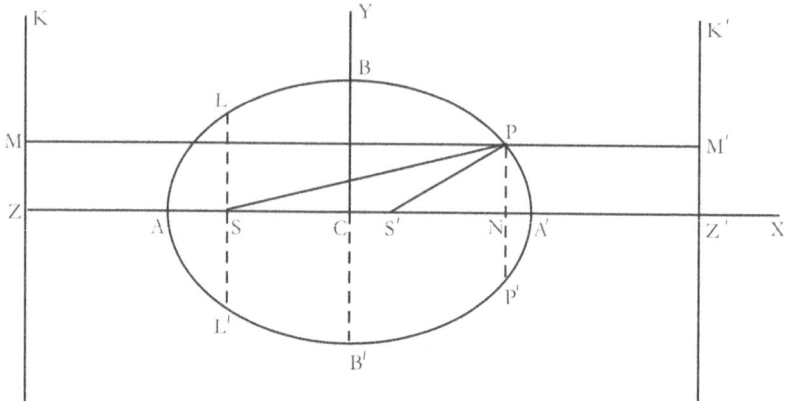

Fig. 120

θ	$\cos\theta$	$\dfrac{l}{\gamma}$	γ	curve
If $\theta = 0$	1	$1 - e$	$\dfrac{l}{1-e}$	point A' in above fig
If $\theta = 0$ to $90°$	Decrease from 1 to 0	Increases from $1 - e$ to 1,	decreases form $\dfrac{l}{1-e}$ to l	portion $A'PBL$
$\theta = 90°$ to $180°$	Decrease from 0 to -1	Increase from 1 to $1 + e$	Decreases from x to $\dfrac{l}{1+e}$	Portion LS where $SA = \dfrac{l}{1+e}$
$\theta = 180°$ to $270°$	Increase from -1 to 0	Decrease from $1 + e$ to 1	Increases $\dfrac{l}{1+e}$ to l	Portion $AL..$
$\theta = 270°$ to $360°$	Increase from 0 to 1	Decrease from 1 to $1 - e$	increase from l to $\dfrac{l}{1-e}$	portion $L'B'P'A'$

Since $\cos(-\theta) = \cos(360 - \theta)$, the curve is symmetrical abort lines SA'

Case 3:

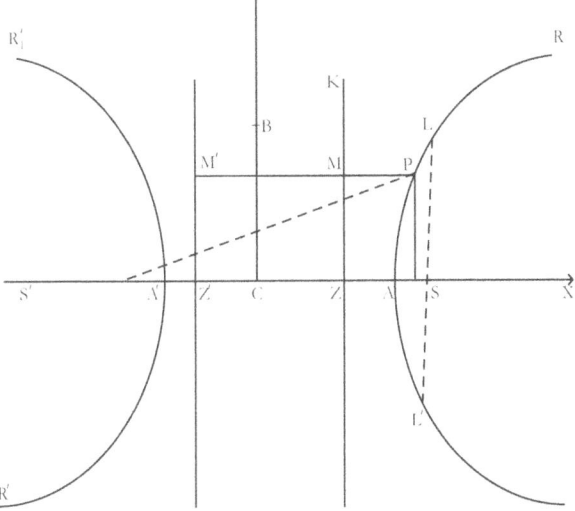

Fig. 121

θ	$\cos\theta$	$\dfrac{l}{\gamma}$	γ	Curve
$\theta = 0$	1	$\begin{aligned} 1-e \\ = -(e-1) \end{aligned}$	$r = \dfrac{-l}{e-1}$	A'
$\theta = 0$ to $\cos^{-1}\left(\dfrac{1}{e}\right)$	Increase from 1 to $\dfrac{1}{e}$	increases $-(e-1)$ to 0	Decreases from $\dfrac{-l}{e-1}$ to $-\infty$	$A'P_1'R'$
$\theta > \cos^{-1}\dfrac{1}{e}$	$\cos\theta < \dfrac{1}{e}$	Small & + ve	Is very great and +ve r changes $-\infty$ to ∞	
θ increase		Increases from 0 to	Decreases	∞RPA

		$1+e$	from ∞ to $\dfrac{l}{1+e}$	
from $\cos^{-1}$ $\dfrac{1}{e}$ to π				
θ increase from π to $2\pi - \cos^{-1}$ $\dfrac{1}{e}$	Increase from -1 to $\dfrac{1}{e}$	Decreases from $1+e$ to 0	Increases from $\dfrac{l}{1+e}$ to ∞	$AL'R_1\infty$
θ increases from $2\pi - \cos^{-1}$ $\dfrac{1}{e}$ to 2π	Increase from $\dfrac{1}{e}$ to 1	Decrease from 0 to $1-e$	γ is negative & decreases from ∞ to $\dfrac{l}{e-1}$	$\infty R_1'A'$

4. Equation to the directrices:

Consider the fig from case (iii), the numerical values of the distances SZ and SZ' are $\dfrac{1}{e}$ and $\dfrac{1}{e}+2c\overline{\chi}$

i.e. $\dfrac{1}{e}$ and $\dfrac{1}{e}+2\dfrac{1}{e(e^2-1)}$

Since $CZ = \dfrac{a}{e} = \dfrac{1}{e(e^2-1)}$

The equations to the two directries are therefore

$$r\cos\theta = \dfrac{-l}{e}$$

And $r\cos\theta = -\left[\dfrac{l}{e}+\dfrac{2l}{e(e^2-1)}\right]$

$$= \dfrac{-l}{e}\left[1+\dfrac{2}{e^2-1}\right]$$

$$= \frac{-l}{e}\left[\frac{e^2 - 1 + 2}{e^2 - 1}\right]$$

$$= \frac{-l}{e} \cdot \frac{e^2 + 1}{e^2 - 1}$$

The same equations would be found to hold in the case of the ellipse.

5. **Equations to the asymptotes:**

The perpendicular distance from S upon an asymptote

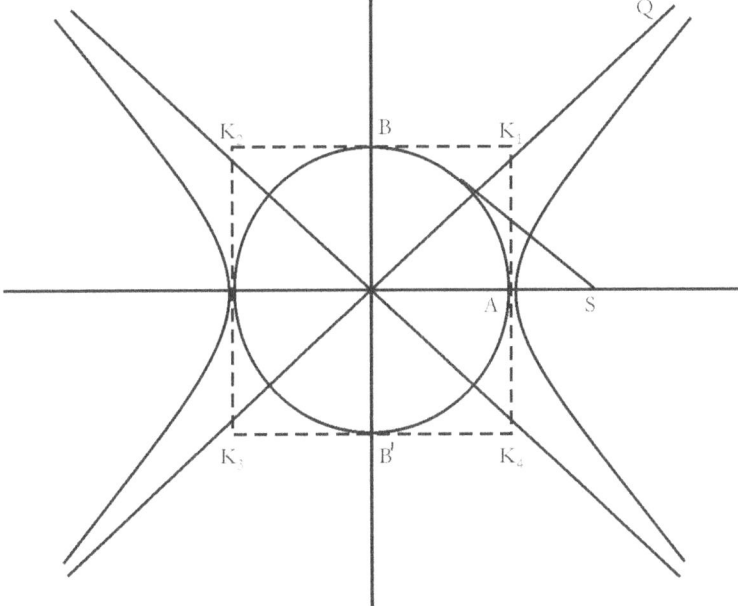

Fig. 122

$$= CS \sin ACK_1$$

$$= ae \cdot \frac{b}{\sqrt{a^2 + b^2}}$$

$$= a \times \frac{\sqrt{a^2 + b^2}}{a} \times \frac{b}{\sqrt{a^2 + b^2}}$$

$$= b$$

$$\because a^2(e^2 - 1) = b^2$$

$$e^2 - 1 = \frac{b^2}{a^2}$$

$$e^2 = \frac{a^2 + b^2}{a^2}$$

$$e = \frac{\sqrt{a^2 + b^2}}{a}$$

Also the asymptote CQ makes an angle $\cos^{-1}\dfrac{1}{e}$ with the axis.

The $\perp^{lar}$ on it from S therefore, makes an angle $\dfrac{\pi}{2} + \cos^{-1}\dfrac{1}{e}$.

Hence the polar equation to the asymptote in general is

$$r\cos(\theta - \alpha)\, p$$

The polar equation to the asymptote CQ is

$$b = r\cos\left[\theta - \frac{\pi}{2} - \cos^{-1}\frac{1}{e}\right]$$

$$= r\cos\left[-\left(\frac{\pi}{2} + \cos^{-1}\frac{1}{e} - \theta\right)\right] \quad (\because \cos(-\theta) = +\cos\theta)$$

$$= r\cos\left[\frac{\pi}{2} - \left(\theta - \cos^{-1}\frac{1}{e}\right)\right] \quad (\because \cos(90 - \theta) = \sin\theta)$$

$$= r\sin\left(\theta - \cos^{-1}\frac{1}{e}\right)$$

The polar equation to the other asymptotes is similarly

$$b = r\cos\left[\theta - \left(\frac{3\pi}{2} - \cos^{-1}\frac{1}{e}\right)\right]$$

$$= r\cos\left[-\left(\frac{3\pi}{2} - \cos^{-1}\frac{1}{e} - \theta\right)\right] \quad [\because \cos(-\theta) = +\cos\theta]$$

$$= r\cos\left(\frac{3\pi}{2} - \left(\cos^{-1}\frac{1}{e} + \theta\right)\right) \quad \cos(270 - \theta) = -\sin\theta$$

$$= -r\sin\left(\theta + \cos^{-1}\frac{1}{e}\right)$$

6. **To find the polar equation of the tangent to any point p of the conic section** $\dfrac{l}{r} = 1 - e\cos\theta$

Let P be the point (r_1, α) and let Q be another point on the curve, whose co-ordinates are $(r_2 - \beta)$ so that we have

$$\frac{l}{r_1} = 1 - e\cos\alpha$$

$$(1)$$

And $\dfrac{l}{r_2} = 1 - e\cos\beta \hspace{3cm} (2)$

We have the polar equation of the straight line joining the point whose co-ordinates are $(r_1, \theta_1) \& (r_2, \theta_2)$ is

$$\frac{\sin(\theta_2 - \theta_1)}{r} = \frac{\sin(\theta - \alpha)}{r_2} + \frac{\sin(\theta_2 - \theta)}{r_1}$$

The polar equation of the line PQ is

$$\frac{\sin(\beta - \alpha)}{r} = \frac{\sin(\theta - \alpha)}{r_2} + \frac{\sin(\beta - \theta)}{r_1}$$

By means of equations (1) & (2) this equation becomes

$$\frac{\sin(\beta - \gamma)}{r} = \sin(\theta - \alpha) \times \frac{1 - e\cos\beta}{l} + \sin(\beta - \theta) \times \frac{1 - e\cos\alpha}{l}$$

Or $\dfrac{l}{r}\sin(\beta - \gamma) = \sin(\theta - \alpha)(1 - e\cos\beta) + \sin(\beta - \theta)(1 - e\cos\alpha)$

$$= \{\sin(\theta - \alpha) + \sin(\beta - \theta)\}$$

$$-e\{\sin(\theta - \alpha)\cos\beta + \sin(\beta - \theta)\cos\alpha\}$$

$$= 2\sin\left(\frac{\theta - \alpha + \beta + \theta}{2}\right)\cos\left(\frac{\theta - \alpha - \beta + \theta}{2}\right)$$

$$-e\{(\sin\theta\cos\theta - \cos\theta\sin\alpha)\cos\phi$$

$$+(\sin\beta\cos\theta - \cos\beta\sin\theta)\cos\alpha\}$$

$$= 2\sin\left(\frac{\beta - \alpha}{2}\right)\cos\left(\theta - \frac{\alpha + \beta}{2}\right)$$

$$-e\{\cos\theta(\sin\alpha\cos\beta - \cos\alpha\sin\beta)\} + \sin\theta(0)$$

$$= 2\sin\left(\frac{\beta - \alpha}{2}\right)\cos\left(\theta - \frac{\alpha + \beta}{2}\right) - e\cos\theta\sin(\beta - \alpha)$$

i.e. $\dfrac{l}{r} = \dfrac{2\sin\left(\dfrac{\beta - \gamma}{2}\right)\cos\left(\theta - \dfrac{\alpha + \beta}{2}\right) - e\cos\theta\sin(\beta - \alpha)}{2\sin\left(\dfrac{\beta - \gamma}{2}\right)\cos\left(\dfrac{\beta - \gamma}{2}\right)}$

$$\frac{l}{r} = \sec\left(\frac{\beta - \gamma}{2}\right)\cos\left(\theta - \frac{\alpha + \beta}{2}\right) - 2e\cos\theta \qquad (3)$$

This is the equation to the straight line joining two point, P and Q on the curve whose vectorial angles α and β, are given

If $\beta = \alpha$ then $\dfrac{l}{r} = \cos(\theta - \alpha) - e\cos\theta$ which is the equation to the tangent at the point α

7. **Let us determine the polar equation of the polar of any point**

   (r_1, θ_1) **with respect to the conic section** $\dfrac{l}{r} = 1 - e\cos\theta$

Let the tangent at the point whose vectorial angles are α and β meet in the point (r_1, θ_1)

The co-ordinates r_1 and θ_1 must therefore, satisfy equation

$$\frac{l}{r} = \cos(\theta - \alpha) - e\cos\theta$$

So that

$$\frac{l}{r_1} = \cos(\theta_1 - \alpha) - e\cos\theta_1 \qquad (1)$$

Similarly $\dfrac{l}{r_1} = \cos(\theta_1 - \beta) - e\cos\theta_1 \qquad (2)$

(1) – (2), we have

$$0 = \cos(\theta_1 - \alpha) - \cos(\theta_1 - \beta) - 0$$

Or $\cos(\theta_1 - \alpha) = \cos(\theta_1 - \beta)$

And therefore,

$$\theta_1 - \alpha = -(\theta_1 - \beta) \ [\because \alpha \text{ and } \beta \text{ are not equal}]$$

$$\theta_1 + \theta_1 = \alpha + \beta$$

$$\Rightarrow \theta_1 = \frac{\alpha + \beta}{\alpha} \qquad (3)$$

Substituting this value in (1), we have

$$\frac{1}{r_1} = \cos\left\{\frac{\alpha + \beta}{2} - \alpha\right\} - e\cos\theta_1$$

i.e. $\cos\left(\dfrac{\alpha + \beta - 2\alpha}{2}\right) = \dfrac{l}{r_1} + e\cos\theta_1$

Or $\cos\left(\dfrac{\beta - \alpha}{2}\right) = \dfrac{l}{r_1} + e\cos\theta_1 \qquad (4)$

Also, by equation $\dfrac{l}{r_2} = 1 - e\cos\beta$

The equation of the line joining the points α and β is

$$\frac{l}{r} + e\cos\theta = \sec\left(\frac{\beta - \alpha}{2}\right)\cos\left(\theta - \frac{\alpha + \beta}{2}\right)$$

i.e. $\left(\dfrac{l}{r}+e\cos\theta\right)\cos\left(\dfrac{\beta-\alpha}{2}\right)=\cos\left(\theta-\dfrac{\alpha+\beta}{2}\right)$

i.e. $\left(\dfrac{l}{r}+e\cos\theta\right)\left(\dfrac{l}{r_1}+e\cos\theta_1\right)=\cos(\theta-\theta_1)$ $\qquad$ (5)

$(\because$ from (3) & (4))

This therefore is the required polar equation to the polar of the point (r_1,θ_1)

8. **Let us determine the equation to the normal at the point whose vectorial angle is α**

The equation to the tangent at the point α is

$$\dfrac{l}{r}=\cos(\theta-\alpha)-e\cos\theta$$

In Cartesian co-ordinates

$$x(\cos\alpha-e)+y\sin\alpha=l \qquad (1)$$

Let the equation to the normal be

$$A\cos\theta+B\sin\theta=\dfrac{l}{r} \qquad (2)$$

i.e. $Ax+By=1 \qquad (3)$

Since (1) & (3) are $\perp$, we have

$$A(\cos\alpha-e)+B\sin\alpha=0 \qquad (4)$$

Since, (2) goes through the point $\left(\dfrac{1}{1-e\cos\alpha},\alpha\right)$ we have

$$A\cos\alpha+B\sin\alpha=1-e\cos\alpha$$
$$(5)$$

Solving (4) and (5), we have

$$A=\dfrac{1-e\cos\alpha}{e}, \text{ and } B=\dfrac{(1-e\cos\alpha)(e-\cos\alpha)}{e\sin\alpha}$$

The equation (2) then becomes

$$\sin\alpha\cos\theta+(e-\cos\alpha)\sin\theta=\frac{le\sin\alpha}{r(1-e\cos\alpha)}$$

i.e. $\sin(\theta-\alpha)-e\sin\theta=\dfrac{e\sin\alpha}{1-e\cos\alpha}\cdot\dfrac{1}{r}$

Solved Examples

Example 1: In any conic, prove that (1) The sum of the reciprocals of the segments of any focal chord is constant, and (2) The sum of the reciprocals of two $\perp^{lar}$ focal chords is constant.

Let PSP' be any focal chord and let the vectorial angle of P be α, so that the vectorial angle of P' is $\pi+\alpha$

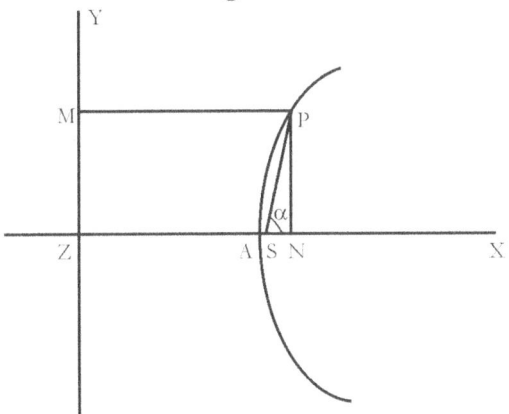

Fig. 123

(1) By equation $r=\dfrac{l}{1-e\cos\alpha}$, we have

$$\frac{l}{SP}=1-e\cos\alpha$$

And $\dfrac{l}{SP'}=1-e\cos(\pi+\alpha)=1+e\cos\alpha$

Hence $\dfrac{l}{SP} + \dfrac{l}{SP'} = 1 - e\cos\alpha + 1 + e\cos\alpha$

$$= 2$$

So that $\dfrac{1}{SP} + \dfrac{1}{SP'} = \dfrac{2}{l}$

The semi-latus-rectum is therefore, the harmonic mean between the segment of any focal chord

(2) Let QSQ' be the focal chord $\perp$ to PSP', so that the vectorial angles of Q and Q' are $\dfrac{\pi}{2} + \alpha$ and $\dfrac{3\pi}{2} + \alpha$, we then have

$$\dfrac{l}{SQ'} = 1 - e\cos\left(\dfrac{\pi}{2} + \alpha\right)$$

$$= 1 + e\cos\alpha$$

And $\dfrac{l}{SQ'} = 1 - e\cos\left(\dfrac{3\pi}{2} + \alpha\right)$

$$(\because \cos(270 + \alpha) = \sin\alpha)$$

$$= 1 - e\sin\alpha$$

Hence $PP' = SP + SP' = \dfrac{l}{1 - e\cos\alpha} + \dfrac{l}{1 + e\cos\alpha}$

$$= \dfrac{l(1 + e\cos\alpha - 1 - e\cos\alpha)}{1^2 - e^2\cos^2\alpha}$$

$$= \dfrac{l(2)}{1 - e^2\cos^2\alpha}$$

And $QQ' + SQ + SQ' = \dfrac{l}{1 + e\sin\alpha} + \dfrac{l}{1 - e\sin\alpha}$

$$= \dfrac{l(1 - e\sin\alpha + 1 + e\sin\alpha)}{1^2 - e^2\sin^2\alpha}$$

$$= \dfrac{2l}{1 - e^2\sin^2\alpha}$$

Therefore $\dfrac{1}{PP'} + \dfrac{1}{QQ'} = \dfrac{1-e^2\cos^2\alpha}{2l} + \dfrac{1-e^2\sin^2\alpha}{2l}$

$$= \dfrac{1-e^2\cos^2\alpha + 1 - e^2\sin^2\alpha}{2l}$$

$$= \dfrac{2-e^2(\cos^2\alpha + \sin^2\alpha)}{2l}$$

$$= \dfrac{2-e^2(1)}{2l}$$

$$\therefore \dfrac{1}{PP'} + \dfrac{1}{QQ'} = \dfrac{2-e^2}{2l}$$

And in therefore, the same for all such pair of chords

Example 2: Prove that the locus of the middle points of focal chords of a conic section is a conic section.

Let PSQ be any chord, the angle PSX being θ

So that $SP = \dfrac{l}{1-e\cos\theta}$

And $SQ = \dfrac{1}{1-e\cos(\pi+\theta)} = \dfrac{1}{1+e\cos\theta}$

Let R be the middle point of PQ, and let its polar. Co-ordinate be r and θ

Then $r = SP - RP$

$$= SP - \dfrac{SP + SQ}{2}$$

$$= \dfrac{2SP - SP - SQ}{2}$$

$$= \dfrac{SP - SQ}{2}$$

$$= \dfrac{1}{2}\left\{ \dfrac{l}{1-e\cos\theta} - \dfrac{l}{1+e\cos\theta} \right\}$$

$$= \frac{l}{2}\left\{\frac{1+e\cos\theta-1+e\cos\theta}{1-e^2\cos^2\theta}\right\}$$

$$r = \frac{l}{2}\left\{\frac{2e\cos\theta}{1-e^2\cos^2\theta}\right\}$$

$$r^2 = \frac{l}{2}\times\frac{2er\cos\theta}{1-e^2\cos^2\theta}$$

$$r^2(1-e^2\cos^2\theta) = ler\cos\theta$$

Or $r^2 - r^2e^2\cos^2\theta = le\cdot r\cos\theta$

Transforming to Cartesian co-ordinates this equation becomes

$$x^2 + y^2 - e^2x^2 = lex$$

If the original conic is parabola then $e = 1$

Then $x^2 + y^2 - x^2 = lx$

$$y^2 = lx$$

If e is not equal to unity, then $x^2 + y^2 - e^2x^2 = lex$ can be written as

$$x^2 - e^2x^2 - lex + y^2 = 0$$

$$x^2(1-e^2)-(1-e^2)\times\frac{lex}{1-e^2}+\frac{1}{4}\frac{l^2e^2}{(1-e^2)^2}(1-e^2)+y^2$$

$$= \frac{l^2e^2}{4(1-e^2)}$$

$$= (1-e^2)\left\{x^2-2\times\frac{1}{2}\frac{le}{1-e^2}x+\left[\frac{1}{2}\frac{le}{(1-e^2)}\right]^2\right\}+y^2 = \frac{l^2e^2}{4(1-e^2)}$$

$$\Rightarrow (1-e^2)\left\{x-\frac{1}{2}\cdot\frac{le}{1-e^2}\right\}^2+y^2 = \frac{l^2e^2}{4(1-e^2)}$$

And therefore represent an ellipse or a hyperbola according as the original conic is an ellipse or a hyperbola.

Example 3: If the tangents at any two point, P and Q of a conic meet in a point T, and if the straight line PQ meet the directrix corresponding to S in a point K, then the angle KST is a right angle.

If the vectorial angles of P and Q be α and β respectively the equation to PQ, is $\dfrac{l}{r} = \sec\left(\dfrac{\beta-\alpha}{2}\right)\cos\left(\theta - \dfrac{\alpha+\beta}{2}\right) - e\cos\theta$ (1)

Also the equation to the directrix is

$$\frac{l}{r} = -e\cos\theta \tag{2}$$

If we solve the equations (1) & (2)

$$-e\cos\theta = \sec\left(\frac{\beta-\alpha}{2}\right)\cos\left(\theta - \frac{\alpha+\beta}{2}\right) - e\cos\theta$$

$$\text{Or} \sec\left(\frac{\beta-\alpha}{2}\right)\cos\left(\theta - \frac{\alpha+\beta}{2}\right) = 0$$

$$\text{Or}\, \theta - \frac{\alpha+\beta}{2} = \frac{\pi}{2}$$

i.e. $\lfloor KSX = \dfrac{\pi}{2} + \dfrac{\alpha+\beta}{2}$

So that SK bisect the exterior angle between SP and SQ

Also by equation $\dfrac{\alpha+\beta}{2} = \theta_1$, we have the vectorial angle of T

equal to $\dfrac{\alpha+\beta}{2}$

i.e. $\lfloor TSX = \dfrac{\alpha+\beta}{2}$

$\therefore \lfloor KST = \lfloor KSX - \lfloor TSX = \dfrac{\pi}{2}$

14. Closing Thoughts

Mathematics is not a spectator sport. The patterns and underlying nuances are like a work of art. The more you apply yourself to the subject, the more you uncover and understand. Mathematics is a subject which requires practice. This is not something that you relax on your couch, casually browse through and hope to achieve mastery. This will require patience and application.

1. Neatness is conducive to accuracy. Refrain from the temptation to write down something quickly and scratching the same to make the necessary corrections.

2. One of the weakness we find in student while solving word problems is the usage of = sign. This sign as a specific meaning in the world of mathematics. It cannot be used as a way to begin every new line of step in the problem solving process. Use appropriate mathematical signs and symbols. Never use them to mean something vague. = Sign is never good space filler.

3. Spend a second or two to explain how you arrived at a certain step. Several books and references use a statement, such as ``it follows from the above statement''. We have oftentimes wondered how the expression or equation below follows from the one above. A good explanation is an excellent demonstration of your understanding of the underlying principles.

4. When you are faced with several conclusions during a problem solving process, it is a good idea to number the statements or equations. In subsequent steps, you can refer to these conclusions by using the label or the assigned equation number.

5. The easiest of problems attracts the silliest of mistakes. If the problem is easy, motivate yourself to get it right. Do not let over-confidence or carelessness to take control of the situation

.

www.ingramcontent.com/pod-product-compliance
Lightning Source LLC
Chambersburg PA
CBHW071402170526
45165CB00001B/151